Lecture Notes
in Economics and
Mathematical Systems

Managing Editors: M. Beckmann and W. Krelle

232

Luc Bauwens

Bayesian Full Information Analysis of Simultaneous Equation Models Using Integration by Monte Carlo

Springer-Verlag
Berlin Heidelberg New York Tokyo 1984

Author

Dr. Luc Bauwens
C.O.R.E.
34, Voie du Roman Pays
B-1348 Louvain-La-Neuve, Belgium

ISBN 3-540-13384-4 Springer-Verlag Berlin Heidelberg New York Tokyo
ISBN 0-387-13384-4 Springer-Verlag New York Heidelberg Berlin Tokyo

Library of Congress Cataloging in Publication Data. Bauwens, Luc, 1952-. Bayesian full information analysis of simultaneous equation models using integration by Monte Carlo. (Lecture notes in economics and mathematical systems; 232) Bibliography: p. 1. Linear models (Statistics). 2. Equations, Simultaneous. 3. Monte Carlo method. I. Title. II. Series.
QA276.B33 1984 519.5 84-14137
ISBN 0-387-13384-4 (U.S.)

Printing and binding: Beltz Offsetdruck, Hemsbach/Bergstr.
2142/3140-543210

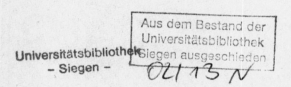

ACKNOWLEDGMENTS

I want to thank Professor J.-F. Richard for his many fruitful suggestions in the course of this research. There is no doubt that this monograph would not have been written if I had not benefited from his enlightening advises during the years I spent at CORE.

My thanks are also due to Professors J. Drèze and H. van Dijk (of Erasmus University at Rotterdam), and to my colleague and friend M. Lubrano, for occasional discussions on the topic.

All these persons, as well as Professors A. Barten and D. Weiserbs, made also many useful comments on the manuscript, and thereby helped me to improve the presentation of this monograph.

E. Pecquereau deserves my gratitude for her prompt and efficient typing.

I am indebted to H. Tompa, although he was not directly concerned with this research, for transmitting and sharing generously his incomparable experience in numerical analysis and computer programming during his long period of activity at CORE.

Financial support of the "Projet d'Action Concertée" of the government of Belgium under contract 80/85-12 is gratefully acknowledged. This contract has also financed the numerous computations I had to make. It has been a pleasure to work on CORE computer.

My wife Bernadette and my children Ariane and Ronald have always comforted me by their agreeable company; let they be praised for their unfailing patience.

TABLE OF CONTENTS

INTRODUCTION

In their review of the "Bayesian analysis of simultaneous equation systems", Drèze and Richard (1983) - hereafter DR - express the following viewpoint about the present state of development of the Bayesian full information analysis of such systems :

i) the method allows "a flexible specification of the prior density, including well defined noninformative prior measures";

ii) it yields "exact finite sample posterior and predictive densities".

However, they call for further developments so that these densities can be evaluated through numerical methods, using an integrated software package. To that end, they recommend the use of a Monte Carlo technique, since van Dijk and Kloek (1980) have demonstrated that "the integrations can be done and how they are done".

In this monograph, we explain how we contribute to achieve the developments suggested by Drèze and Richard.

A basic idea is to use known properties of the porterior density of the parameters of the structural form to design the importance functions, i.e. approximations of the posterior density, that are needed for organizing the integrations. If in particular the prior density is in the class of extended natural-conjugate densities, which has been proposed by Morales (1971) and Drèze and Morales (1976), the main properties we use are the following :

i) the posterior density for the coefficients of a single equation, conditional on the coefficients of the other equations, belongs to the class of 1-1 poly-t densities;

ii) there exist sufficient conditions for the existence of moments of the posterior density of the coefficients of the structural form.

We concentrate on this case, which is reviewed in DR (section 6.3).

In order to give an intuitive view of the usefulness of these properties, let us discuss briefly the method of importance sampling, which is the Monte Carlo technique of numerical integration used throughout this monograph. Let $\kappa(\theta)$ be a kernel of the posterior density of the parameters θ $(\in \Theta)$ and let $g(\theta)$ be an integrable function of θ. The expected value of $g(\theta)$ can be written as

$$(1) \qquad \frac{\int_{\Theta} g(\theta)\, w(\theta)\, f(\theta)\, d\theta}{\int_{\Theta} w(\theta)\, f(\theta)\, d\theta}$$

where $w(\theta) = \kappa(\theta) / f(\theta)$ is the weight function, and $f(\theta)$ is a so-called importance function, i.e. a density function with the following two properties :

i) $f(\theta)$ must be a good approximation of $\kappa(\theta)$, so that $w(\theta)$ varies as little as possible;

ii) it must be possible to generate random drawings from $f(\theta)$.

It is then possible to estimate (1) by a ratio of two sample averages

$$\frac{N^{-1} \sum_{k=1}^{N} g(\theta_k) w(\theta_k)}{N^{-1} \sum_{k=1}^{N} w(\theta_k)} \tag{2}$$

where $\{\theta_k, k = 1, 2, \ldots, N\}$ is a sample of N independent random drawings from $f(\theta)$.

The knowledge of the conditional densities of the posterior can be exploited in several ways to build an importance function; for example, the importance function can be defined as the product of the posterior conditional 1-1 poly-t density for the coefficients of a single equation, and a suitably chosen marginal density for the other coefficients. Also, if an upper bound on the order of existence of moments of the posterior can be imposed upon the importance function, the two density functions are likely to have similar tails; this can prevent the occurence of extreme values (relative to the average) of the weight function $w(\theta)$ when θ is drawn in the tails of the posterior.

The use of 1-1 poly-t densities as (conditional) importance functions requires the availability of an efficient algorithm for generating random drawings from such distributions. Such an algorithm has been proposed by Bauwens and Richard (1982).

An important criterion that has guided this research is a requirement for automatized procedures. The importance functions we propose are *automatic* in the sense that their parameters are the *same* functions of the parameters of the posterior density *whatever the model*. Automatized procedures are needed for implementing a user friendly computer package, thanks to which interested econometricians could estimate a simultaneous equation model in the Bayesian framework, without programming effort. However, such procedures must be carefully thought out, since generality could entail inefficiency; this is particularly true for Monte Carlo methods, where it is important to use as much as possible all the information one has about a problem. We consider therefore several classes of importance functions which are suited to specific cases, so that for a particular model, it should be relatively easy to identify to which case it corresponds.

CONTENTS

The book contains five chapters and four appendices.

Chapter I introduces the notations of the model, discusses its interpretation and presents its likelihood function.

Chapter II considers Bayesian inference with an extended natural-conjugate prior density; two reformulations of the likelihood function are used in order to integrate analytically different subsets of the parameters of the structural form and the section concludes with a discussion of the Monte Carlo technique (importance sampling) that is used to integrate numerically the remaining parameters.

Chapter III proposes a selection of importance functions for the two approaches distinguished in the previous chapter.

Chapter IV reports some practical experiments and draws some general conclusions.

Chapter V indicates various extensions.

Appendix A gives the notations used for the following probability density functions : the matricvariate normal, the multivariate Student, the inverted-Wishart, the 2-0 poly-t and the m-1 poly-t. It discusses also the use of some properties of these distributions for generating random drawings from them.

Appendix B contains some technicalities of Chapter III.

Appendix C contains a few plots of posterior marginal densities and of the corresponding importance functions, for some parameters of the models used in Chapter IV.

Appendix D describes briefly the computer program that has been written to implement the methods described in this monograph.

CHAPTER I : THE STATISTICAL MODEL

I.1. *NOTATION*

We adopt the following formulation of the structural form of the dynamic linear simultaneous equation model :

$$XA = \tilde{Y}B + Z\Gamma = \tilde{U} \tag{1.1}$$

where $X = [\tilde{Y} \quad Z]$;

$\tilde{Y} = [Y \quad Y_I]$ is a $T \times \tilde{m}$ matrix of observed endogenous variables; the submatrix Y is of dimensions $T \times m$;

Z is a $T \times n$ matrix of observed predetermined variables;

$A = \begin{bmatrix} B \\ \Gamma \end{bmatrix}$ is an $(\tilde{m} + n) \times \tilde{m}$ matrix of constants and unknown distinct coefficients; B is a nonsingular matrix of order $\tilde{m}$, whose diagonal elements β_{ii}, for $i = 1,\dots,m$, are equal to 1 (normalisation rule)[1];

$\tilde{U} = [U \quad 0]$ is a $T \times \tilde{m}$ matrix; U is a $T \times m$ matrix of unobserved disturbances; the last $\tilde{m} - m$ columns of $\tilde{U}$ are equal to zero, in which case the corresponding columns of A do not contain unknown parameters (linear identities).

The following assumptions are also made :

i) predeterminedness : $\forall t \leqslant T$, $\forall i \geqslant 0$ such that $t + i \leqslant T$, the row vectors z_t' and u_{t+i}' are linearly independent;

ii) normality : U has a matricvariate normal density (as defined in Appendix A), i.e.

$$p(U) = f_{MN}^{T\times m}(U \mid 0, \Sigma \otimes I_T). \tag{1.2}$$

The i-th equation ($\forall i \leqslant m$) of the system (1.1) can be written as

$$y_i = X_{(i)} \delta_i + u_i \tag{1.3}$$

where y_i is the i-th column of Y;

$X_{(i)} = -[Y_{(i)} \quad Z_{(i)}]$ is the $T \times \ell_i$ submatrix of X consisting of the observations on the ℓ_i explanatory variables appearing in the i-th equation;

δ_i is the $\ell_i \times 1$ vector of unrestricted coefficients in the i-th equation;

u_i is the i-th column of U.

The matrix A contains $\ell = \sum_{i=1}^{m} \ell_i$ unrestricted parameters. Let $\delta = (\delta_1', \ldots, \delta_m')'$ be the $\ell \times 1$ vector of these coefficients.

The reduced form of (1.1) is

$$(1.4) \qquad Y = Z\Pi + V$$

where $\Pi = -\Gamma B^{-1}$ is an $n \times \tilde{m}$ matrix depending on δ ;

$V = \tilde{U} B^{-1} = U B^E$ where B^E is the selection of the first m rows of B^{-1}.

It follows from the above assumption about the density of U that V has also a matricvariate normal density :

$$(1.5) \qquad p(V) = f_{MN}^{T \times \tilde{m}} (V \mid 0, \Omega \otimes I_T)$$

where $\Omega = (B^E)' \Sigma B^E$.

The notation $A(\delta)$, $B(\delta)$, $\Pi(\delta)$ ... will be used whenever the dependence of these matrices on δ is to be emphasized explicitly.

I.2. *INTERPRETATION*

Behavioural equations, endogenous variables and identities

Apart from the identities, the linear relationships in (1.1) are generally thought of as (approximations of) the behaviour of economic agents. Their precise specification is often the result of a preliminary search, in particular as regards the dynamic aspects, the choice of proxy variables, etc.

The endogenous variables $\tilde{y}_t'$ (whose realisations form a typical row of $\tilde{Y}$) are observable random variables representing usually the outcomes of the behaviour of the agents during a period t of fixed length. They are the variables the model builder is interested to model for whatever reasons : e.g. he wants to predict their (expected conditional) values in future periods, or he has to take, or more modestly recommend, decisions based on some characteristics of the underlying behaviours, such as functions of δ .

The identities are also derived from theoretical considerations. Often they represent the coherency conditions to which the outcomes of the acts of the agents are submitted; see Spanos (1982), for a challenging discussion of the role of identities in econometric models. Notice that the identities in (1.1) are restricted to be linear. Simple nonlinearities (in the variables) can be taken into consideration, as will be illustrated in Chapter V.

Predetermined variables

The variables $z_t' \in \mathbb{R}^n$ (whose realisations form a typical row of Z) consist of weakly exogenous variables - see Engle et al. (1983) -, say $w_t \in \mathbb{R}^k$ and lagged values of w_t and $\tilde{y}_t$. The model is conditional on z_t, under the maintained hypothesis that this entails no loss of information relevant for the decision problems of the model builder. Variables whose weak exogeneity is doubtful could be included in $\tilde{y}_t$, e.g. for the purpose of testing their exogeneity, but this typically complicates the analysis and results in a less parsimonious model.

The inclusion of lagged variables in z_t allows for representing dynamic aspects of behaviours, such as partial adjustments, adaptive expectations, mechanisms for correcting past discrepancies.

Completeness of the system

As it is formulated, the structural form (1.1) is a complete system in the sense that there are as many endogenous variables as equations. However, following Richard (1979), the system can also be interpreted as consisting of a number, say p, of structural equations smaller than $\tilde{m}$, with the remaining q $(= \tilde{m} - p)$ equations being unconstrained reduced form equations. In this case, the restrictions, if any, on Π come from the p restricted structural equations linking simultaneously (through B) several endogenous variables and from the identities; similarly the restrictions on Ω come from the same p structural equations and from the restrictions (if any) on the covariance matrix of the disturbances of these equations. Notice that when (1.3) corresponds to one of the q unconstrained reduced form equations, $X_{(i)} = -Z$, δ_i is a column of Π or minus a corresponding column of Γ, while the corresponding column of B is the appropriate column of an identity matrix of order $\tilde{m}$. In the limit case where $p = 0$ and $m = \tilde{m}$, (1.1) is the traditional multivariate regression model, see Zellner (1971, Ch. 8); if in addition exclusion restrictions are allowed in Γ, (1.1) is a restricted traditional multivariate regression model, which can also be viewed as a seemingly unrelated regression model.

I.3. *LIKELIHOOD FUNCTION*

In order to specify the likelihood function, one must make a choice of parameterisation. In the structural form, the parameters are $\delta \in \mathbb{R}^{\ell}$ and $\Sigma \in C_m$ (the set of PDS matrices of order m) both assumed otherwise unrestricted. In the reduced form, they are $\Pi \in P$ and $\Omega \in C_{\tilde{m}}$, where

(1.6) $$P = \{\Pi : \Pi \in \mathbb{R}^{n\tilde{m}} \text{ such that } \exists\, \delta \in \mathbb{R}^{\ell} \text{ for which } \Pi B(\delta) + \Gamma(\delta) = 0\}.$$

If there exist no distinct values of δ, say δ_a and δ_b, for which $\Gamma(\delta_a) \cdot B^{-1}(\delta_a) = \Gamma(\delta_b) \cdot B^{-1}(\delta_b)$, δ is identified (just-identified if $P = \mathbb{R}^{n\tilde{m}}$, and overidentified if P is a strict subset of $\mathbb{R}^{n\tilde{m}}$). If there exist distinct values δ_a and δ_b for which $\Gamma(\delta_a) \cdot B^{-1}(\delta_a) = \Gamma(\delta_b) \cdot B^{-1}(\delta_b)$, δ is not identified. Identification of some components of δ and lack of identification of other components of δ are not mutually exclusive.

In a classical framework a lack of identification typically results in the non-estimability of some parameters. In a Bayesian framework, where sample information is complemented by prior information, the posterior distribution may still be integrable even though δ is not identified, provided the prior density is informative in the directions where the sample information is deficient. For more details, see DR (section 3).

For reasons that will become evident throughout Chapter II, we shall use the parameterisation in terms of δ and Σ and therefore write the likelihood function as

$$(1.7) \qquad L(\delta, \Sigma \mid X) \propto \|B(\delta)\|^T \, |\Sigma|^{-\frac{1}{2}T} \exp -\frac{1}{2} \operatorname{tr} \Sigma^{-1} A_E'(\delta) X'X A_E(\delta)$$

where $A_E(\delta)$ is the selection of the first m columns of $A(\delta)$. Notice that the only factor affected by the presence of identities is $\|B(\delta)\|$, see Rothenberg and Leenders (1964). It is therefore not necessary to solve (1.1) explicitly for its identities, which might induce undesirable reparameterisations of the structural form. Initial conditions are assumed to be known, which amounts to reduce the sample size, and are present in the first row of Z; for a more general treatment, see Richard (1979).

CHAPTER II : BAYESIAN INFERENCE : THE EXTENDED NATURAL-CONJUGATE APPROACH

II.1. *TWO REFORMULATIONS OF THE LIKELIHOOD FUNCTION*

Morales (1971) discusses two reformulations of (1.7) which will prove useful for our purpose. Both are based on the notation (1.3) of structural equations and affect the exponential argument of (1.7) whose dependence on δ is made clear. The first one is named : $AI(\Sigma)$ (for analytical integration of Σ), and the second one : $AI(\gamma)$ (for analytical integration of γ, the subvector of δ regrouping the coefficients of the predetermined variables in the equations (1.3)).

The $AI(\Sigma)$ approach

Defining the matrices

$$\Delta = \begin{bmatrix} \delta_1 & 0 & \dots & 0 \\ 0 & \delta_2 & \dots & 0 \\ \vdots & \vdots & \ddots & \vdots \\ 0 & 0 & \dots & \delta_m \end{bmatrix}, \text{ of dimensions } \ell \times m,$$

and

$$\Xi = [X_{(1)} \quad X_{(2)} \dots X_{(m)}], \text{ of dimensions } T \times \ell,$$

the exponential argument of (1.7) is equivalent to

$$(2.1) \qquad -\frac{1}{2} \operatorname{tr} \Sigma^{-1} (Y - \Xi\Delta)'(Y - \Xi\Delta)$$

$$= -\frac{1}{2} \operatorname{tr} \Sigma^{-1} [S + (\Delta - \hat{\Delta})' M (\Delta - \hat{\Delta})]$$

where $M = \Xi'\Xi$

$\hat{\Delta} = M^{+}\Xi' Y$ (the superscript + is used to denote a generalised inverse)

$S = (Y - \Xi\hat{\Delta})'(Y - \Xi\hat{\Delta})$.

The matrix M is singular as soon as some of the $X_{(i)}$ have columns in common. If columns of Y are present in Ξ, S is also a singular matrix. See e.g. DR (section 6.3). It is assumed that the matrix $(Y - \Xi\Delta)'(Y - \Xi\Delta)$, which is equal to $A_E' X'X A_E$, is almost surely of rank m.

The AI(γ) approach

Defining the matrix

$$x = \begin{bmatrix} X_{(1)} & 0 & \cdots & 0 \\ 0 & X_{(2)} & \cdots & 0 \\ \vdots & \vdots & \ddots & \vdots \\ 0 & 0 & \cdots & X_{(m)} \end{bmatrix}, \text{ of dimensions } mT \times \ell,$$

and

$y = \text{vec } Y$, the column expansion of Y, of dimensions $mT \times 1$,

the exponential argument of (1.7) is also equivalent to

$$(2.2) \qquad -\frac{1}{2}(y - x\delta)'(\Sigma^{-1} \otimes I_T)(y - x\delta)$$

$$= -\frac{1}{2}(\delta - \hat{\delta})'(\Sigma^{-1} \square M)(\delta - \hat{\delta})$$

where M is the $\ell \times \ell$ matrix $\Xi'\Xi$ with typical block $M_{ij} = X'_{(i)} X_{(j)}$, which we denote $M = [M_{ij}]$

$\Sigma^{-1} \square M \underset{\text{def}}{=} [\sigma^{ij} M_{ij}]$, an $\ell \times \ell$ matrix with typical block $\sigma^{ij} M_{ij}$ of dimensions $\ell_i \times \ell_j$; so that $\Sigma^{-1} \square M = x'(\Sigma^{-1} \otimes I_T)x$

$\hat{\delta} = (\Sigma^{-1} \square M)^{-1} x'(\Sigma^{-1} \otimes I_T)\, y$

$= (\Sigma^{-1} \square M)^{-1} [\Sigma^{-1} \square \Xi'(i'_m \otimes I_T)]\, y$ (i_m is an m vector of ones).

This reformulation has been used by Rothenberg and Leenders (1964) in the classical framework and by Morales (1971), Drèze and Morales (1976) and Richard (1973) in the Bayesian framework; see also DR (section 6.5).

II.2. THE EXTENDED NATURAL-CONJUGATE PRIOR DENSITY

Drèze and Morales (1976) have defined an extended natural-conjugate prior density on the basis of (2.1) :

$$(2.3) \qquad p(\delta, \Sigma) \propto \| B \|^{\tau_0} |\Sigma|^{-\frac{1}{2}(\nu_0 + m + 1)} \exp -\frac{1}{2} \text{tr } \Sigma^{-1} [S_0 + (\Delta - \Delta_0)' M_0 (\Delta - \Delta_0)],$$

where S_0 (of order m) and M_0 (of order ℓ) are PSDS matrices. A necessary condition for the integrability of (2.3) is that $S_0 + (\Delta - \Delta_0)' M_0 (\Delta - \Delta_0)$ is almost surely PDS; this condition is verified if S_0 and M_0 are PDS, as will be assumed further (except otherwise indicated).

The special case, where

(i) $\tau_0 = 0$,

(ii) $\nu_0 > m-1$ and $\nu_0 > \mu = \sup\{\ell_i, i = 1,\ldots,m\}$,

(iii) $S_0 = \text{diag}(s_{01},\ldots, s_{0m})$, with $s_{0i} > 0$, $i = 1,\ldots,m$,

(iv) M_0 is block diagonal with nonsingular diagonal blocks M_{0ii}, $i = 1,\ldots,m$,

(v) Δ_0 has the same "column-diagonal" structure as Δ - we denote by δ_{0i} the subvector of Δ_0 corresponding to δ_i in Δ -

is easy to interpret since under these conditions, $p(\delta)$ factorizes into a product of independent Student densities with a common exponent, i.e.

$$(2.4) \qquad p(\delta) = \prod_{i=1}^{m} f_t^{\ell_i}(\delta_i \mid \delta_{0i}, s_{0i}^{-1} M_{0ii}, \nu_0 - \ell_i),$$

(see Appendix A for notations of density functions).

The main limitation of the marginal prior density (2.4) is the built-in independence between any pair δ_i, δ_j. It is therefore especially convenient when the prior information on δ comes naturally in this form, e.g. if it originates from independent (possibly hypothetical) samples pertaining to each equation, as was illustrated by Morales (1971, Part 2). The common exponent restriction is not very serious : if ν_0 is large enough to ensure the existence of at least second order moments of every δ_i, i.e. if $\nu_0 - \mu > 2$, M_{0ii} or s_{0i} can be adjusted to obtain the required prior covariance matrix of δ_i, say V_{0ii}. As this matrix is given by

$$(2.5) \qquad V_{0ii} = \frac{s_{0i}}{\nu_0 - \ell_i - 2} M_{0ii}^{-1},$$

it is clear that many values of s_{0i}, M_{0ii} and ν_0 are compatible with a given value of V_{0ii}. The selection of a particular value of these parameters is crucial for the weight that the prior information has relative to the sample information; in particular, the value of M_{0ii} must be compared with the value of M_{ii}, the corresponding sample matrix, to check its relative weight. One must also check that the prior marginal expectation of Σ, if it exists, remains a priori sensible (see below). These difficulties arise because δ and Σ are not a priori independent. Note also that the experiments of Morales (1971, Part 2) have led him to conclude in favor of a large value of ν_0 in this class of prior densities.

Let us make a few additional comments about the specification of the prior density (2.3) :

1. There is a noninformative prior on δ and Σ which is obtained from (2.3) for $\tau_0 = 0$, $S_0 = 0$ and $M_0 = 0$:

$$(2.6) \qquad p(\delta, \Sigma) \propto |\Sigma|^{-\frac{1}{2}(\nu_0+m+1)}$$

where ν_0 is often chosen in application of some invariance principle. Zellner (1971) recommends setting ν_0 equal to 0, in partial application of Jeffreys' invariance principles, Drèze (1976) recommends setting $\nu_0 = n$, on the basis of an invariance argument specific to the case of the simultaneous equation model, and Malinvaud (1978, p. 268) recommends setting $\nu_0 = m-1$ on the basis of an invariance argument specific to the case of the traditional multivariate regression model. In any case the choice of ν_0 is not simple to the extent that it must ensure at least the integrability of the posterior density of δ and that it affects the existence and the order of magnitude of the posterior moments of δ.

The prior density (2.3) can be partially noninformative on some parameters; e.g. if $S_0 = \text{diag}\,(0 \ldots s_{0i} \ldots 0)$, $s_{0i} > 0$, and M_{0ii} is PDS but $M_{0jj} = 0$, $\forall j \neq i$, while $\tau_0 = 0$, $\nu_0 > m-1$ and $\nu_0 > \ell_i$, the prior density is noninformative on all the parameters except σ^{ii} and δ_i, whose marginal density is the same as in (2.4). In such a case, it is obvious that (2.3) is not integrable in the space of δ and Σ, i.e. $\mathbb{R}^\ell \times C_m$. One can also be noninformative on some elements of δ_i, by setting the corresponding rows and columns of M_{0ii} equal to 0.

2. If Δ_0 has the same structure as Δ, (2.3) can be written so as to combine naturally with (1.7) modified by (2.2) :

$$(2.7) \qquad p(\delta, \Sigma) \propto \|B\|^{\tau_0} \, |\Sigma|^{-\frac{1}{2}(\nu_0+m+1)} \exp -\frac{1}{2}\left[\operatorname{tr} \Sigma^{-1} S_0 + (\delta-\delta_0)'(\Sigma^{-1} \,\square\, M_0)(\delta-\delta_0)\right]$$

where $\delta_0 = (\delta'_{01}, \ldots, \delta'_{0m})'$.

If Δ_0 has not the same[2] structure as Δ, $\delta_0'(\Sigma^{-1} \,\square\, M_0)\delta_0$ must be subtracted from the scalar between the square brackets of (2.7), and δ_0 must be defined in the same way as $\hat{\delta}$ in (2.2), i.e.

$$(2.8) \qquad \delta_0 = (\Sigma^{-1} \,\square\, \Xi_0' \Xi_0)^{-1} \left[\Sigma^{-1} \,\square\, \Xi_0' (i_m' \otimes I_{\tau_0})\right] y_0$$

where Ξ_0 is any $\tau_0 \times \ell$ matrix such that $M_0 = \Xi_0' \Xi_0$

$y_0 = \text{vec}\,(Y_0)$; Y_0 is any $\tau_0 \times m$ matrix such that $M_0 \Delta_0 = \Xi_0' Y_0$ and

$S_0 = (Y_0 - \Xi_0' \Delta_0)' (Y_0 - \Xi' \Delta_0)$.

In other words, given a previous (hypothetical) sample size τ_0, we compute Y_0 and Ξ_0 from the given matrices M_0, Δ_0 and S_0 of (2.3), by solving relations similar to those stated after (2.1); δ_0 is then computed from Y_0 and Ξ_0 in the same way as $\hat{\delta}$ is computed from Y and Ξ.

3. If $\nu_0 > m-1$, the marginal density of δ is

(2.9) $$p(\delta) \propto \|B\|^{\tau_0} \left|S_0 + (\Delta - \Delta_0)' M_0 (\Delta - \Delta_0)\right|^{-\frac{1}{2}\nu_0}$$

This density is in general asymmetrical, but it is symmetrical under less restrictive conditions on the parameters than the conditions (i) - (v) stated above.

PROPOSITION 1 : If $\tau_0 = 0$ *or* $|B| = 1$ *identically in* δ *, and* $\nu_0 > m-1$*, a necessary and sufficient condition for* $p(\delta)$ *to be symmetrical around its mode* $\delta_0 = (\delta'_{01}, \ldots, \delta'_{0m})'$ *is that* Δ_0 *has the same "column-diagonal" structure as* Δ*, i.e.*

(2.10) $$\Delta_0 = \begin{bmatrix} \delta_{01} & 0 & \cdots & 0 \\ 0 & \delta_{02} & \cdots & 0 \\ \vdots & \vdots & \ddots & \vdots \\ 0 & 0 & \cdots & \delta_{0m} \end{bmatrix}.$$

Proof : We use the following definition : $p(\delta)$ is symmetrical around δ_0 if and only if for any distinct values δ_a and δ_b for which $\delta_a - \delta_0 = \delta_0 - \delta_b$, i.e. $\delta_0 = \frac{1}{2}(\delta_a + \delta_b)$, $p(\delta_a) = p(\delta_b)$.

1 - necessity : if $p(\delta)$ is symmetrical, there exist two values δ_a and δ_b for which $\delta_0 = \frac{1}{2}(\delta_a + \delta_b)$; defining Δ_a and Δ_b from δ_a and δ_b respectively, as δ from Δ, the previous equality implies that $\Delta_0 = \frac{1}{2}(\Delta_a + \Delta_b)$: this shows that Δ_0 must be "column-diagonal";

2 - sufficiency : if Δ_0 is "column-diagonal", for any δ_a, define the corresponding "column-diagonal" Δ_a; let $\Delta_b = 2\Delta_a - \Delta_0$, which defines δ_b ; δ_a and δ_b satisfy the relations $\delta_a - \delta_0 = \delta_0 - \delta_b$ and $p(\delta_a) = p(\delta_b)$ (since $\Delta_a - \Delta_0 = \Delta_0 - \Delta_b$).

It is immediate that δ_0 is the mode of $p(\delta)$, i.e. $p(\delta_0) > p(\delta)$, $\forall \delta \neq \delta_0$, for $|S_0|^{-\frac{1}{2}\nu_0} > |S_0 + (\Delta - \Delta_0)' M_0 (\Delta - \Delta_0)|^{-\frac{1}{2}\nu_0}$, as follows from the assumption that S_0 and M_0 are PDS.

In any case, it is clear that if $\|B\|^{\tau_0}$ depends on some elements of δ, (2.9) may be asymmetrical. □

A straightforward corollary of this proposition is that δ_0 is also the prior expectation of δ.

Assuming again that $\tau_0 = 0$ or $\|B\| = 1$, it is easily seen from (2.7) that the conditional density $p(\delta \mid \Sigma)$ is a normal density; if Δ_0 is "column-diagonal", the conditional expectation (or mode) does not depend on Σ and coincides with the marginal expectation (or mode) δ_0; if Δ_0 is not "column-diagonal", this is no longer true, though there are probably many values of Δ_0 for which (2.10) does not hold and the marginal density (2.9) is moderately asymmetrical. Further investigations are needed to trace the influence of relaxing the zero restrictions of (2.10) on the symmetry of (2.9) when $\tau_0 = 0$ or $\|B\| = 1$.

Obviously if these conditions are not verified, the factor $\|B\|^{\tau_0}$ may introduce much skewness in (2.9). However, if we except the case where the prior information genuinely originates from a previous sample, we do not think that it would be appropriate to specify a skewed prior by selecting a positive τ_0 (the natural choice, since τ_0 is the sample size of a hypothetical sample) : indeed the skewness would come mainly from the structure of B, being more or less exacerbated according to the value of τ_0, and the prior would depend critically on the specification of the model.

4. If $\nu_0 > m-1$, the conditional density of Σ is

$$p(\Sigma \mid \delta) = f_{iW}^{m}(\Sigma \mid S_0 + (\Delta - \Delta_0)' M_0 (\Delta - \Delta_0), \nu_0). \tag{2.11}$$

Though the marginal density $p(\Sigma)$ cannot be obtained analytically,[3] the marginal expectation of Σ can be obtained from the expectation and the covariance matrix of δ :

<u>*PROPOSITION 2*</u> *: If* $\nu_0 > m+1+\mu+\tau_0$, *and* M_{0ii}, $i = 1,\dots,m$ *are PDS matrices,*

$$E(\Sigma) = \frac{1}{\nu_0 - m - 1}\left[S_0 + \Delta_0' M_0 \Delta_0 - \Delta_0' M_0 E(\Delta) - E(\Delta') M_0 \Delta_0 + E(\Delta' M_0 \Delta)\right] \tag{2.12}$$

where $E(\Delta)$ is defined from $E(\delta)$ as Δ from δ, and

$$E(\Delta' M_0 \Delta) = [\mathrm{tr}\, M_{0ij}\, \mathrm{Cov}(\delta_i, \delta_j) + E(\delta_i') M_{0ij} E(\delta_j)]. \tag{2.13}$$

<u>*Proof*</u> *:* From (2.11), if $\nu_0 > m+1$,

$$E(\Sigma \mid \delta) = \frac{1}{\nu_0 - m - 1}\left[S_0 + (\Delta - \Delta_0)' M_0 (\Delta - \Delta_0)\right]. \tag{2.14}$$

We have then simply to apply the well known formula

$$E(\Sigma) = E_\delta[E(\Sigma \mid \delta)]$$

to obtain (2.12), taking account of lemma 6.6 of DR, wherefrom the stated conditions on ν_0 and M_{0ii}, $i = 1,\dots,m$, are sufficient for the existence of the first two moments of δ.

Finally, (2.13) follows from

$$E(\Delta' M_0 \Delta) = [E(\delta_i' M_{0ij} \delta_j)]$$

and the well known general result

$$E[(z-m)'Q(z-m)] = \operatorname{tr} QV + (\mu - m)'Q(\mu - m),$$

for $z \in \mathbb{R}^\ell$, $E(z) = \mu$, $V(z) = V$, and constant $m \in \mathbb{R}^\ell$ and Q of order ℓ. □

5. If any of the restrictions (i) - (v) is relaxed, as in some previous remarks, (2.9) must be integrated numerically in order to assess its informative content. Fo example, one may wish to maintain (i) - (iii) and (v), but to introduce probabilistic dependence between δ_i and δ_j through a non zero off-diagonal block M_{0ij} of M_0.

   As we shall advocate in section II.5 for the posterior marginal density, a Monte Carlo procedure is required. When $p(\delta)$ is symmetrical or even moderately skew, we shall see from the experiments of Chapter IV that this can be performed at very reasonable cost in an automatic way.

Other properties of the density (2.3) are stated in subsection III.2.1., but except for the first one, they are not very useful for the elicitation of prior information.

II.3 *POSTERIOR DENSITIES*

The AI(Σ) approach

By combining (1.7), modified by (2.1), with (2.3), the posterior density of δ and Σ is seen to have the same form as (2.3), i.e.

$$(2.15) \qquad p(\delta, \Sigma \mid X) \propto \|B\|^{\tau_*} |\Sigma|^{-\frac{1}{2}(\nu_* + m + 1)} \exp -\frac{1}{2} \operatorname{tr} \Sigma^{-1} [S_* + (\Delta - \Delta_*)' M_* (\Delta - \Delta_*)]$$

where
$$\tau_* = \tau_0 + T$$
$$\nu_* = \nu_0 + T$$
$$M_* = M_0 + M$$
$$\Delta_* = M_*^+ (M_0 \Delta_0 + M \hat{\Delta})$$
$$S_* = S_0 + \Delta_0' M_0 \Delta_0 + S + \hat{\Delta}' M \hat{\Delta} - \Delta_*' M_* \Delta_*,$$

wherefrom (2.9) and (2.11) are the corresponding posterior densities upon substituting the subscript $*$ for the subscript 0 ; see DR (section 6.3).

The AI(γ) approach

From (1.7), (2.2) and (2.7) the joint posterior density can be put in the form - see DR (section 6.5) -

(2.16) $$p(\delta, \Sigma \mid X) \propto \|B\|^{\tau_*} |\Sigma|^{-\frac{1}{2}(\nu_*+m+1)} \exp -\frac{1}{2}\left[\text{tr}\, \Sigma^{-1} S_0 - \delta_*'(\Sigma^{-1} \square M_*)\delta_* + (\delta - \delta_*)'(\Sigma^{-1} \square M_*)(\delta - \delta_*)\right],$$

where

$$\delta_* = (\Sigma^{-1} \square M_*)^{-1}\left[(\Sigma^{-1} \square M_0)\,\delta_0 + (\Sigma^{-1} \square M)\,\hat{\delta}\,\right].$$

This form of the posterior density is useful to obtain the conditional density

(2.17) $$p(\delta \mid \Sigma, X) \propto \|B\|^{\tau_*} f_N^{\ell}(\delta \mid \delta_*, [\Sigma^{-1} \square M_*]^{-1}).$$

Denoting by β the $(\ell_\beta \times 1)$ subvector of δ regrouping the coefficients of the endogenous variables in the equations (1.3), since $B = B(\delta) \equiv B(\beta)$, it follows that

(2.18) $$p(\gamma \mid \beta, \Sigma, X) = f_N^{\ell_\gamma}(\gamma \mid \gamma_* - [\Sigma^{-1} \square M_{*\gamma\beta}][\Sigma^{-1} \square M_{*\beta\beta}]^{-1}[\beta - \beta_*], [\Sigma^{-1} \square M_{*\gamma\gamma}]^{-1})$$

where γ_* and β_* are selected from δ_* as γ and β from δ, M_* being partitioned accordingly into $M_{*\gamma\gamma}$, $M_{*\beta\beta}$, $M_{*\gamma\beta}$, $M_{*\beta\gamma}$; moreover

(2.19) $$p(\beta \mid \Sigma, X) \propto \|B\|^{\tau_*} f_N^{\ell_\beta}(\beta \mid \beta_*, V_{\beta\beta})$$

where

(2.20) $$V_{\beta\beta} = [(\Sigma^{-1} \square M_{*\beta\beta}) - (\Sigma^{-1} \square M_{*\beta\gamma})(\Sigma^{-1} \square M_{*\gamma\gamma})^{-1}(\Sigma^{-1} \square M_{*\gamma\beta})]^{-1}.$$

In Chapter III, we shall discuss how some properties of these densities can be exploited to define the importance functions which are required in order to obtain posterior moments.

II.4 PREDICTIVE MOMENTS

The likelihood function (1.7) is the usual reinterpretation of the joint density of the T rows of $\tilde{Y}$ given the initial conditions, the T observations of the exogenous variables and the parameters. However, when the model is dynamic, the right-hand side of (1.7) is not a convenient formulation of the density of $\tilde{Y}$ because some observations in $\tilde{Y}$ are present also in Z. In fact, following Richard (1979), the data density can be factorized as

(2.21) $$\prod_{t=1}^{T} p(\tilde{y}_t \mid z_t, \Pi, \Omega)$$

where $p(\tilde{y}_t \mid z_t, \Pi, \Omega) = f_N^{\tilde{m}}(\tilde{y}_t \mid \Pi' z_t, \Omega)$. Here we use the parameters Π and Ω, but remember that $\Pi = \Pi(\delta)$ and Ω are the functions of δ and Σ defined at the end of section I.1, so that implicitly the parameterisation remains in terms of δ and Σ.

Computing the first and second order predictive moments for the period $T+1$ requires simply to know the posterior expected values of Π and Ω, since

(2.22) $$E(\tilde{y}_{T+1} \mid w_{T+1}, X, \Pi, \Omega) = \Pi' z_{t+1}$$

(2.23) $$V(\tilde{y}_{T+1} \mid w_{T+1}, X, \Pi, \Omega) = \Omega .$$

Computing multiperiod predictive moments would require to know the posterior moments of higher order of Π, Ω and functions thereof; see e.g. van Dijk and Kloek - hereafter VDK - (1977) for more details. Chow (1973) has treated the case where Π is unrestricted.

II.5 *NUMERICAL INTEGRATION BY IMPORTANCE SAMPLING*

We have to cope with the problem of computing useful characteristics of the posterior distribution of δ and Σ. In the absence of a specific loss function, these characteristics are usually : expected values, a covariance matrix, skewness coefficients, plots of univariate marginal densities and distribution functions, plots of contours of bivariate marginal densities, of the parameters and functions thereof. Such functions typically include :

- impact, interim and total multipliers, needed to evaluate the effects of various policy scenarios; see e.g. Theil (1971, section 9.7);
- the expected values of future observations of the endogenous variables $(\tilde{y}_{T+1}, \tilde{y}_{T+2}, \ldots)$ conditional on values of exogenous variables $(w_{T+1}, w_{T+2}, \ldots)$ and on the observed sample (X), see e.g. (2.22) for one step ahead "predictions";
- the dominant latent root of the characteristic equation of the reduced form written as a system of linear difference equations, and its modulus, since these quantities provide information on the dynamic properties of the system, which can be damped or explosive and oscillatory or monotone.

All these functions can be considered explicitly as functions of the structural coefficients δ and possibly of Σ. Linear functions of Ω can be marginalised with respect to δ taking advantage of (2.14). For example,

(2.24) $$E(P\Omega R \mid \delta) = \frac{1}{\nu_* - m - 1} P(B^E)' [S_* + (\Delta - \Delta_*)' M_* (\Delta - \Delta_*)] B^E R,$$ i.e. a function of δ.

In many cases, all the integrations can thus be carried out with respect to δ. However, if moments of order 2 or univariate marginal densities of elements of Σ or Ω are requested, the numerical integrations have to be performed also with respect to Σ.

We denote by $\theta \in \Theta$ the parameters with respect to which the numerical integrations have to be done, i.e. θ is a subset of δ and Σ. Let $g(\theta)$ be an integrable function of θ; let also $\kappa(\theta)$ be a kernel of $p(\theta)$, the (prior or posterior) density of θ, with integrating constant K^{-1}, i.e.

$$(2.25) \qquad p(\theta) \propto \kappa(\theta), \int_\Theta \kappa(\theta)\, d\theta = K.$$

We wish to compute μ_g, the expected value of $g(\theta)$, i.e.

$$(2.26) \qquad \mu_g = \frac{1}{K} \int_\Theta g(\theta)\kappa(\theta)\, d\theta = \frac{\int_\Theta g(\theta)\kappa(\theta)\, d\theta}{\int_\Theta \kappa(\theta)\, d\theta} .$$

If the dimensions of θ is "large", standard methods of numerical integration (e.g. product rules) are not competitive with Monte Carlo methods; see e.g. Davis and Rabinowitz (1975, section 5.10). As was illustrated by Kloek and van Dijk (1978), hereafter KVD, one may resort in particular to the Monte Carlo technique of importance sampling. Let us briefly review this technique in general terms.

As usual with Monte Carlo methods, one expresses the quantity to be computed as the expected value of a random variable with finite variance. In (2.26), there are two quantities to be computed, namely the numerator and the denominator of the ratio of the right-hand side. Since the latter is a particular case - $g(\theta) = 1$ - of the former, we consider the computation of the former, and express it as

$$(2.27) \qquad \int_\Theta g(\theta)\kappa(\theta)\, d\theta = \int_\Theta g(\theta)\, w(\theta)\, f(\theta)\, d\theta$$

i.e. as the expected value μ_{gw} [4] of $g(\theta)w(\theta)$ with respect to a density $f(\theta)$. The function $w(\theta)$ is the *weight function*, defined as

$$(2.28) \qquad w(\theta) = \kappa(\theta) / f(\theta).$$

The density $f(\theta)$ is the *importance function*. It must be chosen in a class of density functions wherefrom random samples can be efficiently drawn. Let $\{\theta_k, k=1,\dots,N\}$ be such a sample of size N.

Using the so-called moment method of estimation theory - see e.g. Cramer (1946, p. 497), we can estimate μ_{gw} by the sample mean

(2.29) $$m_{gw} = \frac{1}{N} \sum_{k=1}^{N} g(\theta_k)\, w(\theta_k)$$

and $\mu_g = \mu_{gw} / \mu_w$ by the ratio of sample means m_{gw} / m_w, denoted hereafter H for simplicity.

As we wish naturally to control the precision of this estimator, we need a bound on its relative error.

<u>*DEFINITION*</u> : The percentage relative error of an estimator H of a parameter μ_g $(\neq 0)$ is

(2.30) $$\varepsilon = 100 \left| \frac{H}{\mu_g} - 1 \right| .$$

A maximum relative error of x% is obtained if $\varepsilon < \frac{1}{2} x$ $(\forall \mu_g \neq 0)$.

Since we are using a statistical approach to estimation, we can at best make a probabilistic statement about ε. By application of a "central-limit" theorem stated in Cramer (1946, p. 366), H is asymptotically normally distributed with mean μ_g and variance σ^2/N ; by application of formula (27.7.3) of Cramer, under suitable regularity conditions, σ^2 is approximately given by

(2.31) $$\sigma^2 = \frac{1}{\mu_w^2} \sigma_{gw}^2 + \frac{\mu_{gw}^2}{\mu_w^4} \sigma_w^2 - 2 \frac{\mu_{gw}}{\mu_w^3} \operatorname{cov}(gw, w)$$

(in this formula, σ_{gw}^2 and σ_w^2 are the population variances of the random variables $g(\theta)\, w(\theta)$ and $w(\theta)$, θ having the density $f(\theta)$; cov (gw, w) is the corresponding covariance).

Therefrom, the event

(2.32) $$\left\{ \frac{|H - \mu_g|}{\sigma / \sqrt{N}} < z_{\frac{1}{2}\alpha} \right\} = \left\{ \varepsilon < \frac{2}{\sqrt{N}}\, z_{\frac{1}{2}\alpha} \frac{\sigma}{\mu_g} \cdot 100 \right\}$$

where $z_{\frac{1}{2}\alpha}$ is the $1 - \frac{1}{2}\alpha$ fractile of the standard normal distribution, is of probability $1 - \alpha$, at least, for large N. In other words, it can then be asserted with level of confidence $1 - \alpha$ that the relative error committed with H is less than an upper bound $\bar{\varepsilon}$ given by

(2.33) $$\bar{\varepsilon} = (2 / \sqrt{N})\, z_{\frac{1}{2}\alpha}\, (\sigma / \mu_g) \cdot 100.$$

It is seen that this bound depends on two factors which may be controlled more or less easily :

(i) it depends inversely on $\sqrt{N}$, so that increasing the sample size reduces in principle $\overline{\varepsilon}$, but at a slow rate; e.g. if we increase N by a factor of 100, we increase the accuracy of the answer by a factor of 10 only; see e.g. Davis and Rabinowitz (1975, Ch. 5);

(ii) $\overline{\varepsilon}$ depends directly on σ/μ_g ; it is illuminating to express the square of this ratio, taking account of (2.31) and of the relation $\mu_g = \mu_{gw}/\mu_w$, as was done by KVD :

$$(2.34) \qquad \frac{\sigma^2}{\mu_g^2} = \frac{\sigma_{gw}^2}{\mu_{gw}^2} + \frac{\sigma_w^2}{\mu_w^2} - 2\,\rho_{gw,w}\,\frac{\sigma_{gw}}{\mu_{gw}}\,\frac{\sigma_w}{\mu_w}$$

($\rho_{gw,w}$ is the simple correlation coefficient between gw and w).

All the population moments [illegible] 34) are of course unknown in practice and are therefore replaced [illegible]nding sample moments; $\overline{\varepsilon}$ is then an *estimated* relative e[illegible]

[illegible] first two terms are the squared coeffi[illegible]ables gw and w. The last term will de[illegible] w and μ_{gw} have the same sign, the [illegible] may be expected if the variance of g [illegible]ase, g almost never changes sign [illegible] does gw whose sample mean keeps [illegible]ariance of g means that the varia[illegible] in which case these two random [illegible]y, if g has a small variance, [illegible]iation of gw and of w, i.e. [illegible] 34) be of the same order of ma[illegible]

Evi[illegible] requires at least that the ratio [illegible]tion of gw and of w are almost [illegible] so that σ_w^2 be as small as poss[illegible]hoose f so that σ_{gw}^2 be as sma[illegible] the beginning of this section, w[illegible]f several functions of θ, i.e. g([illegible]entative function. Trying to ch[illegible]cular function g could be time [illegible] necessary to draw a random sample [illegible] concludes in favor of selecting the importan[illegible] σ_w^2 is the smallest possible ("variance reduction"), i.e. [illegible]ght function is the most constant; the density f should "mimic" [illegible] p, but this may be difficult to achieve since the properties of p are rarely well known.

It may be worth to mention that *uniform* convergence of the estimators m_w and m_{gw}/m_w to their respective limits K and μ_g is not guaranteed by the well known properties according to which these estimators converge in probability, and with probability 1 to their limits, as $N \rightarrow \infty$. If f is a poor approximation of p, it may even be difficult to obtain reliable estimates of K $(=\mu_w)$ and σ_w^2 : for all practical sample sizes, m_w does not seem to converge to a stable value, while s_w^2 (the sample variance of w) increases rather discontinuously, being influenced occasionally by extreme values of w. It may however happen at the same time that estimates of expected values of several functions g and of their variances seem to converge to stable values, which are however *incorrect*[5]; this is especially the case for the variances. Therefore, an apparent lack of convergence of m_w or s_w^2 for a reasonable value of N is a serious warning about the quality of the importance function. These remarks do not contradict the convergence properties mentioned above, since these results are established for $N \rightarrow \infty$.

CHAPTER III : SELECTION OF IMPORTANCE FUNCTIONS

III.1. *GENERAL CRITERIA*

As indicated at the end of section II.5, the choice of the functional form and of the parameters of an importance function $f(\theta)$ for a density $p(\theta)$ should be guided by the requirement that f be a good approximation of p. The following criteria are in our opinion necessary but by no means sufficient conditions to attain this goal.

i) A criterion of "*location*" : f should be located in the same region of Θ as p, so as to concentrate the most important part of its probability in the same region as p. This can be partially obtained if p and f have the same modes or expected values.

ii) A criterion of "*variance-covariance structure*" : the covariance matrix of θ under f should be a fairly good approximation of the true, but of course unknown, covariance matrix. At a simple level this covers two aspects : the size of each variance, and the correlation structure, i.e. the value of each simple correlation coefficient. For example, an incorrect correlation structure will induce too many drawings in regions where the ratio p/f is great or low, relative to its value for drawings located near the mode, leading to unreliable estimates of the integral and sometimes even of the moments of the posterior density. A more elaborate way to state this criterion of "variance-covariance structure" is to require that the (largest) eigenvalues and the corresponding eigenvectors of the covariance matrix of f be approximately equal to those of the covariance matrix of p, since a covariance matrix V can be decomposed as $Q \Lambda Q'$, where Λ is the diagonal matrix of eigenvalues of V and Q is the orthogonal matrix of the corresponding eigenvectors.

iii) A criterion of "*behaviour of the tails*" : the tails of f should not diverge too much from those of p, to prevent again extreme values of p/f. It is well known that the behaviour of the tails of a density is closely linked to the order of existence of its moments; therefore f should be characterized by the same order of existence of its moments as p.

There is of course a certain interdependence between these criteria. For example, if f has much thinner tails than p, because f is taken to be normal and has therefore moments of any order, it will usually be characterized by smaller variances than p.

If p is very skewed in some directions, even if the two density functions have the same ellipsoids of concentration, then clearly f may be a poor approximation, since its level hypersurfaces may look like far from being homothetic to those of p; this is obvious if f is taken to be symmetrical. This example illustrates that the three criteria stated above may not be sufficient to guarantee that the importance function be good. Ideally of course, f should not only have moments up to the same order as p, but also the same moments as p, although in many cases, almost equality of moments of order 1 to 3 might be sufficient.

III.2. *THE* $AI(\Sigma)$ *APPROACH*

This approach requires the construction of an importance function for δ. The importance functions we propose are based on analytical properties of the posterior density of δ. We begin by stating these properties through four lemmas (subsection III.2.1), drawing upon the review of DR (section 6.3). We consider then (subsection III.2.2) a class of Student importance functions, which has already been used by VDK (1980), and three classes of poly-t based importance functions :

i) a product of independent 1 - 1 poly-t densities for the coefficients δ_i of each equation (1.3); each 1 - 1 poly-t density is taken to be the posterior density of δ_i, conditional on $\delta_{\bar{i}} = (\delta_1' \delta_2', \ldots, \delta_{i-1}' \delta_{i+1}', \ldots, \delta_m')'$, the coefficients of all equations except the i-th; in other words, each marginal importance function for δ_i is a conditional density $p(\delta_i \mid \delta_{\bar{i}}, X)$ of the posterior $p(\delta \mid X)$, and we fix the conditioning values as the modal values of $p(\delta \mid X)$; see subsection III.2.3;

ii) a mixture of density functions; each member of the mixture is defined as the importance function of the previous case, except that the conditioning values differ for each member; the mixing distribution is of course a distribution of the conditioning values, and we take account of the fact that the vectors $\delta_{\bar{1}}, \delta_{\bar{2}}, \ldots, \delta_{\bar{m}}$ have common subvectors, i.e. are defined from a common random vector $\phi (= \delta)$; a Student mixing density is proposed in subsection III.2.4;

iii) a conditional 1 - 1 poly-t density $p(\delta_i \mid \delta_{\bar{i}}, X)$, for a chosen value of i, and a suitably defined marginal importance function for $\delta_{\bar{i}}$, say $f(\delta_{\bar{i}})$; we propose to select $f(\delta_{\bar{i}})$ in the Student family, but we also discuss whether it can be decomposed into conditional and marginal densities; see subsection III.2.5.

In the following subsections, we emphasize the main ideas and we refer therefore the reader to Appendix B for technical details. The last subsection (III.2.6) points out the automatic aspect of the importance functions proposed.

III.2.1. *Properties of the posterior density of* δ

A first lemma gives a sufficient condition for the existence of moments of δ, whose marginal density is obtained from (2.15) as

$$(3.1) \qquad p(\delta \mid X) \propto \|B\|^{\tau_*} \, |S_* + (\Delta - \Delta_*)' M_* (\Delta - \Delta_*)|^{-\frac{1}{2}\nu_*}.$$

The parameters τ_*, ν_*, S_* $(m \times m)$, M_* $(\ell \times \ell)$ and Δ_* $(\ell \times m)$ are defined after formula (2.15). We recall that $\mu = \sup_i \{\ell_i, i = 1,\ldots,m\}$.

LEMMA 1 : Let $p(\delta)$ *be the density* (3.1), *with* S_* *and* M_{*ii}, $i = 1,\ldots,m$, PDS *matrices. A sufficient condition for the existence of moments of order* r *is*

$$(3.2) \qquad \nu_* - m + 1 - \mu - \tau_* > r.$$

Proof. See DR (lemma 6.6).

Note that, since $\nu_* = \nu_0 + T$ and $\tau_* = \tau_0 + T$, (3.2) is also the corresponding condition for the prior density (2.9). In other words, the sample size T does not affect the order of existence of moments of the posterior density. The condition that M_{*ii} be PDS excludes an exact multicollinearity in the matrix $X_{(i)}$ of (1.3) if $M_{0ii} = 0$; more generally, it means that the prior must be informative in the directions where the sample information is deficient if there is a linear dependence between the columns of $X_{(i)}$. Note however that it is not assumed that M_* be PDS; it need only be PSDS. Finally, the condition that S_* be PDS means also that prior information must be provided at least on some variances of the disturbances of the structural equations; indeed if the prior is noninformative as (2.6), S_* is equal to the matrix S of (2.1), a singular matrix if columns of Y are present in Ξ.

The following lemma gives the functional form of the conditional posterior density $p(\delta_i \mid \delta_{\bar{i}})$ [6]. We recall that $\delta_{\bar{i}}$ is the $\ell - \ell_i$ subvector obtained by deleting δ_i from δ.

LEMMA 2 : Let $p(\delta)$ *be the density* (3.1), *with* S_* *and* M_{*ii} PDS *matrices. The conditional density* $p(\delta_i \mid \delta_{\bar{i}})$ *of* $p(\delta)$ *is a* $1-1$ *poly-t density, i.e.*

$$(3.3) \qquad p(\delta_i \mid \delta_{\bar{i}}) \propto \varphi_{1,1}(\delta_i \mid \delta_{\bar{i}}) = Q_{0i}^{\lambda_0} \, Q_{1i}^{-\lambda_1}$$

where Q_{0i} (≥ 0) *and* Q_{1i} (> 0) *are quadratic forms in* δ, *i.e.*

$$Q_{.i} = \bar{s}_{.i} + (\delta_i - \bar{y}_{.i})' \bar{A}_{.i} (\delta_i - \bar{y}_{.i}), \text{ and}$$

$$\lambda_0 = \frac{1}{2}\tau_*, \quad \lambda_1 = \frac{1}{2}\nu_*.$$

The parameters of the two quadratic forms are functions of the parameters S_*, M_* *and* Δ_* *of* $p(\delta)$ *and of the value of the conditioning variables* $\delta_{\bar{i}}$; *see Appendix* B.I *for* *a precise definition.*

Proof. See Appendix B.I. □

We know analytically the kernel of $p(\delta)$, see (3.1), and the density $p(\delta_i \mid \delta_{\bar{i}})$, see (3.3); consequently, we know the kernel of the marginal density.

LEMMA 3 : Under the conditions of lemma 2, the marginal density $p(\delta_{\bar{i}})$ *of* $p(\delta)$ *is given by*

(3.4) $$p(\delta_{\bar{i}}) \propto k(\delta_{\bar{i}})\, p_i(\delta_{\bar{i}})$$

where

(3.5) $$k(\delta_{\bar{i}}) = \int_{\mathbb{R}^{\ell_i}} \varphi_{1,1}(\delta_i \mid \delta_{\bar{i}})\, d\,\delta_i,$$ *i.e. the reciprocal of the integrating constant of* $p(\delta_i \mid \delta_{\bar{i}})$

(3.6) $$p_i(\delta_{\bar{i}}) = |B_{\bar{i}}' B_{\bar{i}}|^{\frac{1}{2}\tau_*}\, |\bar{S}_i + (\Delta_{\bar{i}} - \bar{\Delta}_i)' \bar{M}_i (\Delta_{\bar{i}} - \bar{\Delta}_i)|^{-\frac{1}{2}\nu_*}.$$

$B_{\bar{i}}$ *is obtained from* B *by deleting the i-th column;*

$\Delta_{\bar{i}}$ *is obtained from* Δ *by deleting the i-th column and the* ℓ_i *corresponding rows;*

$\bar{S}_i$, $\bar{\Delta}_i$, $\bar{M}_i$ *are functions of* S_*, Δ_*, M_* *defined in* Appendix B.I.

Proof. The presence of $k(\delta_{\bar{i}})$ in the kernel of $p(\delta_{\bar{i}})$ is trivial, while the presence of $p_i(\delta_{\bar{i}})$ is justified in Appendix B.I. □

Finally, it is possible to decompose recursively the kernel of (3.1) into a sequence of functions which are defined in the following lemma. We denote this kernel $\kappa(\delta)$, i.e. $\kappa(\delta)$ is equal to the right-hand side of (3.1).

LEMMA 4 : Under the conditions of lemma 1, the kernel $\kappa(\delta)$ *of* $p(\delta \mid X)$ *can be factorized into a product of functions, as follows*

(3.7) $$\kappa(\delta) = \varphi_{1,1}(\delta_1 \mid \delta_{\bar{1}})\, p_1(\delta_{\bar{1}}) = \varphi_{1,1}(\delta_1 \mid \delta_2 \ldots \delta_m)\, \varphi_{1,1}(\delta_2 ; \delta_3 \ldots \delta_m)\, \varphi_{1,1}(\delta_3 ; \delta_4 \ldots \delta_m) \ldots \varphi_{1,1}(\delta_m).$$

where $\varphi_{1,1}(\delta_1 \mid \delta_2 \ldots \delta_m)$ *is defined by* (3.3) *adapted to the case* $i = 1$,

$\varphi_{1,1}(\delta_2 ; \delta_3 \ldots \delta_m)$ *is obtained from* $p_i(\delta_{\bar{i}})$, *i.e.* (3.6) *adapted to the case i = 1, in the same way as* $\varphi_{1,1}(\delta_1 \mid \delta_2 \ldots \delta_m)$ *is obtained from* $\kappa(\delta)$*; it can therefore be expressed as a ratio of two quadratic forms in* δ_2 *(raised to the powers* λ_0 *and* λ_1*), whose parameters are functions of* $\delta_3 \ldots \delta_m$*,*

and the other functions up to $\varphi_{1,1}(\delta_m)$ *are defined by a recursive application of the procedure outlined for* $\varphi_{1,1}(\delta_2 ; \delta_3 \ldots \delta_m)$.

Proof. As shown in Appendix B.I, $\kappa(\delta) = \varphi_{1,1}(\delta_1 \mid \delta_{\bar{1}})\, p_1(\delta_{\bar{1}})$; from (3.6), it can be seen that $p_1(\delta_{\bar{1}})$ depends on $\delta_{\bar{1}}$ in the same way as $\kappa(\delta)$ depends on δ; therefore an application of the procedures described in Appendix B.I to the function $p_1(\delta_{\bar{1}})$ allows to decompose it into a product of $\varphi_{1,1}(\delta_2 ; \delta_3 \ldots \delta_m)$ and a function which depends on $\delta_3 \ldots \delta_m$ in the same way as $\kappa(\delta)$ depends on δ; a complete decomposition (3.7) is obtained by a recursive application of this technique. □

There are obviously $m!$ decompositions of $\kappa(\delta)$ as given in the right-hand side of (3.7). Attention is drawn to the fact that (3.7) is not a decomposition of $p(\delta)$ into a sequence of densities (see lemma 3).

III.2.2. *Student importance function* (STUD)

We reconsider the Student importance function proposed by VDK (1980), say

$$(3.8) \qquad f(\delta) = f_t^{\ell}(\delta \mid d, P, \lambda).$$

In order to satisfy as best as possible the 3 criteria stated in section III.1, the parameters are chosen in the following way :

- d (the expected value of the Student density) as the posterior mode;
- P so that the covariance matrix of (3.8), i.e., $P^{-1}/(\lambda - 2)$, is equal to minus the inverse of the Hessian matrix of $\log p(\delta)$, evaluated at the posterior mode; in this way one hopes to capture the main features of the true covariance matrix, provided the sample size is large enough to validate the use of an asymptotic approximation of the kind considered here;
- $\lambda = \nu_* - m + 1 - \mu - \tau_*$, since λ is the order of existence of moments of (3.8), while the latter value is known to ensure the existence of moments of $p(\delta)$ up to that order, by lemma 1.

The major shortcoming of (3.8) is its symmetry, while $p(\delta)$ can be strongly asymmetrical. If this is the case, no choice of parameters can make (3.8) a good importance function, since in particular the variance-covariance structure given by the stated choice of P might be very distorted.

The computation of d and of the Hessian matrix of the log posterior is discussed briefly in Appendix B.II. The generation of random drawings from a Student density is discussed in Appendix A.

III.2.3. *Poly-t based importance function : case I* (PTFC)

An application of lemma 2 to each coefficient vector δ_i suggests the use of the following importance function :

$$(3.9) \qquad f(\delta) = \prod_{i=1}^{m} p(\delta_i \mid \delta_{\bar{i}} = d_{\bar{i}})$$

where $d_{\bar{i}}$ is obtained from d, the posterior mode, as $\delta_{\bar{i}}$ from δ. We call this importance function PTFC (for Poly-t, Fixed Conditions). It is clearly a product of independent densities for each δ_i since the conditioning values are selected once for all.

In order to generate a random drawing from (3.9), one has simply to draw a value δ_i from $p(\delta_i \mid \delta_{\bar{i}} = d_{\bar{i}})$, for each i. An algorithm to obtain random drawings from 1 - 1 poly-t densities is given in Appendix A.

As regards the "location criterion" for (3.9), the mode of that density will be usually close to that of $p(\delta)$, in view of the stated choice of the conditioning values.

The major problem with (3.9) is that it will usually fail to have the correct variance-covariance structure for 2 reasons :

i) the variances will be underestimated because of the conditional nature of the densities $p(\delta_i \mid \delta_{\bar{i}})$;

ii) (3.9) is a product of independent densities, so that the simple correlation coefficients between the parameters of different equations are forced to zero.

This importance function will therefore be good if the conditional poly-t densities are almost insensitive to the values of the conditioning variables, i.e. are close to marginal densities.

As regards the order of existence of moments, by properties of 1 - 1 poly-t densities (see Appendix A), moments of (3.9) exist up to the order r if $r < 2(\lambda_1 - \lambda_0) - \ell_i = \nu_* - \tau_* - \ell_i$; the latter value is larger than $\nu_* - m + 1 - \tau_* - \mu$, the corresponding value for $p(\delta)$. We suggest therefore to decrease the value of λ_1 to

$$(3.10) \qquad \lambda_{1i} = \frac{1}{2}(\nu_* - m + 1 - \mu + \ell_i).$$

Note that decreasing λ_1 will inflate the covariance matrix of (3.9), but will not alter its correlation structure.

III.2.4. *Poly-t based importance function : case II* (PTDC)

With the hope to find a solution to the problem of the incorrect variance-covariance structure of PTFC, consider the following variant : instead of fixing $\delta_{\bar{i}} = d_{\bar{i}}$ once for all, one can vary the value of $\delta_{\bar{i}}$ used for each drawing of (3.9), in a way which introduces dependence in probability between the coefficients of different equations. Such a procedure raises several issues :

- how do we choose the value of $\delta_{\bar{i}}$ $(i = 1,\dots,m)$ to be used at each drawing ?
- what is the precise definition of the importance function ? Notice that at each drawing the ratio κ/f of the kernel of p and f – see (2.29) and (2.28) – must be computed. The computation of f at any given drawing δ requires an explicit definition of this density function;
- is the procedure worthwhile and feasible in practice ?

The choice of a set of conditioning values $\delta_{\bar{i}}$ for each density $p(\delta_i \mid \delta_{\bar{i}})$ of (3.9) can be made at random, according to some probability law. Denote by $\phi \in \mathbb{R}^{\ell}$ a vector of auxiliary random variables to be drawn from a density $h(\phi)$ in a first stage, in order to fix $\delta_{\bar{i}} = \phi_{\bar{i}}$ for drawing δ from (3.9) in a second stage, with $\phi_{\bar{i}}$ selected from ϕ as $\delta_{\bar{i}}$ from δ. That is, we generate a random drawing δ in two steps :

i) we draw ϕ from a density $h(\phi)$ and select the m vectors $\phi_{\bar{i}}$ from ϕ ;

ii) we draw δ_i from the 1 – 1 conditional poly-t $p(\delta_i \mid \delta_{\bar{i}} = \phi_{\bar{i}})$, for each i, as is done for PTFC.

An obvious choice of $h(\phi)$ is the Student density (3.8) approximating $p(\delta)$:

$$(3.11) \qquad h(\phi) = f_t^{\ell}(\phi \mid d, P, \lambda),$$

or a discrete probability density function, whch could be an approximation of (3.11), say

$$(3.12) \qquad h(\phi) = \{w_k = P[\phi = \phi_k],\ k = 1,\dots,L\},$$

where $\phi_1,\dots,\phi_L$ define a grid of $\mathbb{R}^{\ell}$; w_k could be chosen as the probability of a suitable neighbourhood of ϕ_k in (3.11); if a hypercube is chosen as such a neighbourhood, the techniques proposed by Govaerts (1983) can be used to compute w_k, taking care of course that the hypercubes form a partition of $\mathbb{R}^{\ell}$.

Under this approach, the importance function PTDC (for Poly-T, Drawn Conditions) is defined as

$$(3.13) \qquad f(\delta) = \int_{\mathbb{R}^\ell} \prod_{i=1}^{m} p(\delta_i \mid \phi_{\bar{i}}) \cdot h(\phi)\, d\phi$$

i.e. $f(\delta)$ is the marginal density of the joint density of δ and ϕ. When $h(\phi)$ is given by (3.12), it is the finite mixture

$$(3.14) \qquad f(\delta) = \sum_{k=1}^{L} w_k \prod_{i=1}^{m} p(\delta_i \mid \phi_{\bar{i}k}).$$

Note that (3.9) corresponds to a particular case of (3.14), i.e. $L = 1$, $w_1 = 1$, $\phi_{\bar{i}1} = d_{\bar{i}}$.

In order to compute $f(\delta)$ as defined by (3.13) at any given point (typically a random drawing), one has to compute the expected value of the rather awkward function of ϕ, $\prod_{i=1}^{m} p(\delta_i \mid \phi_{\bar{i}})$, with respect to the Student distribution, i.e.

$$(3.15) \qquad f(\delta) = E_\phi \left[\prod_{i=1}^{m} p(\delta_i \mid \phi_{\bar{i}}) \right].$$

Since the parameters of each $p(\delta_i \mid \phi_{\bar{i}})$ as well as their integrating constants - which have to be taken into account (this is not the case with (3.9)) - are very complicated functions of $\phi_{\bar{i}}$ (described in Appendix B.I), (3.15) has to be computed numerically. We explain in Appendix B.III how this can be organized efficiently for each drawing of δ generated by the two step procedure described above.

The considerations in Appendix B.III suggest that the procedure is feasible in practice if the number of coefficients (ℓ) is not too large. Whether it is worthwhile is a question which we cannot answer in theory. In particular, even if it is clear that the importance functions (3.13) or (3.14) do not impose independence in probability between all the pairs δ_i, δ_j, as was the case with PTFC, it cannot be proved that they are characterized by the correct form of dependence and variance-covariance structure. Some practical experiments are reported in Chapter IV.

The procedure described above is a particular and extreme application of a technique called "conditional Monte Carlo" by Hammersley and Handscomb (1979, Ch. 6) [7].

III.2.5. *Poly-t based importance function : case III* (PTST)

In spite of the rather pessimistic comments made about PTFC and PTDC, especially as regards their variance-covariance structure, these importance functions have still the theoretical advantage that they can be asymmetrical. Since they are based on conditional densities of $p(\delta)$, in which the factor $\|B\|^{\frac{1}{2}T^*}$ is mainly responsible for the skewness, they should retain to some extent that skewness in the directions where it occurs. We propose now an importance function which is based on the knowledge of the conditional density (3.3) and of the marginal density (3.4). As a complete characterization of the dependence between random variables is given by the decomposition of their joint density into a conditional density and a marginal density, we may thus be able to control better the probabilistic dependence between δ_i and $\delta_{\bar{i}}$, for a chosen value of i.

Since

$$(3.16) \qquad p(\delta) = p(\delta_i \mid \delta_{\bar{i}}) \; p(\delta_{\bar{i}})$$

and we can generate random drawings from $p(\delta_i | \delta_{\bar{i}})$, for any value $\delta_{\bar{i}}$, we can use this conditional density in the importance function. We define the latter by

$$(3.17) \qquad f(\delta) = p(\delta_i \mid \delta_{\bar{i}}) \; f(\delta_{\bar{i}})$$

where $f(\delta_{\bar{i}})$ is an importance function for $p(\delta_{\bar{i}})$. We must of course find a good approximation of $p(\delta_{\bar{i}})$ and select the index i of the equation for which we use the conditional 1 - 1 poly-t density in the importance function.

The problem of choice of an importance function $f(\delta_{\bar{i}})$ for the marginal density $p(\delta_{\bar{i}})$ is similar to the original problem of choice of an importance function $f(\delta)$ for $p(\delta)$. But there are two differences :

i) the dimension is reduced from ℓ to $\ell - \ell_i$, which is often an advantage in Monte Carlo techniques;

ii) more importantly, the density $p(\delta_{\bar{i}})$ cannot in general be decomposed usefully in a way similar to (3.16), but we defer a discussion of this topic to the end of this subsection.

Let us first suggest the following importance function for $p(\delta_{\bar{i}})$:

$$(3.18) \qquad f(\delta_{\bar{i}}) = f_t^{\ell-\ell_i}(\delta_{\bar{i}} \mid \bar{d}_i, \bar{P}_i, \bar{\lambda}_i)$$

i.e. a Student density. The parameters of (3.18) could be defined from $p(\delta_{\bar{i}})$ in the same way as the parameters of (3.8) are defined from $p(\delta)$. Another choice of parameters would be obtained by taking (3.18) as the marginal density of (3.8).

The importance function defined by (3.17) - (3.18) will be referred to as PTST-i (for Poly-T-Student, i being the sequence number of the equation corresponding to δ_i). In order to get a random drawing from this importance function, we have simply to draw $\delta_{\bar{i}}$ from (3.18) and δ_i from $p(\delta_i \mid \delta_{\bar{i}})$ where $\delta_{\bar{i}}$ is taken to be the value drawn from (3.18).

The best choice of i is in principle the one for which the variance of the ratio

$\kappa(\delta)\,/\,f(\delta)$ of the kernel of $p(\delta)$ and $f(\delta)$ is the smallest. For PTST-i, due to the presence of an exact conditional importance function $p(\delta_i \mid \delta_{\bar{i}})$, this ratio can be written as

$$(3.19) \qquad \frac{\kappa(\delta)}{f(\delta)} = \frac{\varphi_{1,1}(\delta_i \mid \delta_{\bar{i}}) \cdot p_i(\delta_{\bar{i}})}{k^{-1}(\delta_{\bar{i}}) \cdot \varphi_{1,1}(\delta_i \mid \delta_{\bar{i}}) \cdot f(\delta_{\bar{i}})} = \frac{k(\delta_{\bar{i}}) \cdot p_i(\delta_{\bar{i}})}{f(\delta_{\bar{i}})} ,$$

i.e. the ratio of the kernel of $p(\delta_{\bar{i}})$ and $f(\delta_{\bar{i}})$. The variance of this ratio can therefore be estimated on the basis of preliminary random drawings from (3.18), for each value of i, provided m is not too large. Other considerations which can guide the choice of i are

- the value of ℓ_i : a large value of ℓ_i may imply a smaller variance of the ratio of the right-hand side of (3.19), but on the other hand, it means a larger computer time to generate a random drawing of the importance function;
- the structure of the correlation matrix of (3.8); if it happens that for some value of i, the elements of δ_i are almost uncorrelated with those of $\delta_{\bar{i}}$, that value of i should not be considered in first place; δ_i and $\delta_{\bar{i}}$ may then be almost linearly independent; such a dependence can be incorporated easily in the Student importance function for $\delta_{\bar{i}}$ and it seems then preferable to use the conditional poly-t (with another choice of i) in order to capture a more complicated form of probabilistic dependence. It may also happen that δ_i and $\delta_{\bar{i}}$ are linearly independent but are at the same time very much dependent in probability, in which case the corresponding $p(\delta_i \mid \delta_{\bar{i}})$ could have been a good choice.

These heuristic criteria can be used to choose a particular value, should several be good candidates in view of the results of the preliminary procedure described above, or to eliminate a priori some values of i in this procedure, should m be large.

Moreover, a look at the marginal density $f(\delta_i)$ of PTST-i is interesting :

$$(3.20) \qquad f(\delta_i) = \int_{\mathbb{R}^{\ell_i}} p(\delta_i \mid \delta_{\bar{i}}) \quad f(\delta_{\bar{i}})\, d\,\delta_{\bar{i}} \,.$$

It is straightforward to see that with PTDC as given by (3.13), one obtains also (3.20) as marginal, but for all values of i. With PTST-i, if we change the value of i, we change also $f(\delta_{\bar{i}})$, whereas with PTDC, the weighting function corresponding to $f(\delta_{\bar{i}})$ is the marginal of a common underlying joint density, namely $f(\delta)$ given by (3.8).

Finally, PTST-i seems to be convenient in the case where only δ_i is the parameter of interest and the required results are the expected values of functions $g(\delta_i)$ whose conditional expected values are known analytically. Indeed,

(3.21) $$E[g(\delta_i)] = E_{\delta_{\bar{i}}}[E[g(\delta_i) \mid \delta_{\bar{i}}]]$$

can be estimated by

(3.22) $$\frac{1}{N}\sum_{k=1}^{N} E[g(\delta_i) \mid \delta_{\bar{i}k}] \frac{\kappa(\delta_{\bar{i}k})}{f(\delta_{\bar{i}k})}$$

for a random sample $\delta_{\bar{i}1},\ldots,\delta_{\bar{i}N}$ of the importance function $f(\delta_{\bar{i}})$. It is not necessary in this case to generate random drawings from $p(\delta_i \mid \delta_{\bar{i}})$. Such g functions include the moments and the marginal densities of δ_i. This approach requires of course that a good importance function for $p(\delta_{\bar{i}})$ can be obtained.

If it turns out that the Student approximation (3.18) of $p(\delta_{\bar{i}})$ is not satisfactory, e.g. because $p(\delta_{\bar{i}})$ is far from being symmetrical, other importance functions $f(\delta_{\bar{i}})$ must be designed.

A first approach could consist in factorizing $p(\delta_{\bar{i}})$ into a conditional density for the coefficients δ_j $(j \neq i)$ of one equation, and a marginal density for the remaining coefficients. Assuming for convenience of notations that $i = 1$ and $j = 2$, we can indeed use lemmas 3 and 4 to isolate the conditional density

(3.23) $$p(\delta_2 \mid \delta_3 \ldots \delta_m) \propto k(\delta_2\, \delta_3 \ldots \delta_m)\, \varphi_{1,1}(\delta_2; \delta_3 \ldots \delta_m).$$

Its kernel is the product of the reciprocal of the integrating constant of the density $p(\delta_1 \mid \delta_2\, \delta_3 \ldots \delta_m)$ and of the $1-1$ poly-t function for δ_2 described in lemma 4. Unfortunately the kernel of the marginal density of $\delta_3 \ldots \delta_m$ is known only up to the reciprocal of the integrating constant of (3.23) :

(3.24) $$p(\delta_3 \ldots \delta_m) \propto \int_{\mathbb{R}^{\ell_i}} k(\delta_2\, \delta_3 \ldots \delta_m)\, p_1(\delta_2\, \delta_3 \ldots \delta_m) d\delta_2$$

$$= v(\delta_3\, \delta_4 \ldots \delta_m) \int_{\mathbb{R}^{\ell_i}} k(\delta_2\, \delta_3 \ldots \delta_m)\, \varphi_{1,1}(\delta_2; \delta_3 \ldots \delta_m)\, d\delta_2$$

where $v(\delta_3\, \delta_4 \ldots \delta_m) = \varphi_{1,1}(\delta_3; \delta_4 \ldots \delta_m) \ldots \varphi_{1,1}(\delta_m)$, i.e. the product of the last $m-2$ factors of the right-hand side of (3.7).

Formulas (3.23) and (3.24) suggest to generalize (3.17) into

(3.25) $$f(\delta) = p(\delta_1 \mid \delta_2\, \delta_3 \ldots \delta_m)\, f(\delta_2 \mid \delta_3 \ldots \delta_m)\, f(\delta_3 \ldots \delta_m)$$

where $f(\delta_3 \ldots \delta_m)$ is a Student approximation of (3.24), and $f(\delta_2 \mid \delta_3 \ldots \delta_m)$ is an approximation of (3.23); e.g. if $k(\delta_2\, \delta_3 \ldots \delta_m)$ can be approximated by a Student kernel in δ_2, for given $\delta_3 \ldots \delta_m$, $f(\delta_2 \mid \delta_3 \ldots \delta_m)$ may be defined as a $2-1$ poly-t density. In the particular case where $k(\delta_2\, \delta_3 \ldots \delta_m)$ does not depend on δ_2, or depends very little on δ_2, $f(\delta_2 \mid \delta_3 \ldots \delta_m)$ may be taken as the 1 – 1 poly-t function for δ_2 of the right-hand side of (3.23).

A second approach could be to try to identify the directions of skewness of $p(\delta_{\bar{i}})$ from the results obtained by using PTST-i, and to build then an importance function $f(\delta_{\bar{i}})$ incorporating the same kind of skewness. Such an importance function could be a poly-t density or a quadratic form in normal variables, having the same moments of order 1, 2 and 3 as the posterior (estimated roughly from PTST-i). One has to design procedures to select the parameters of such density functions so that they have the requested moments. Some results in Bauwens and Richard (1982) could be helpful (see in particular their Appendix).

III.2.6. *Conclusion*

It may be worth to point out that the importance functions proposed in this section are automatic in the sense that their parameters can be obtained either as functions of the parameters of the posterior density or through automatized procedures applied to the posterior density.

For example, the parameters of STUD result essentially from the application of a standard technique of maximisation of a function. In the case of PTFC, the parameters of each conditional 1 – 1 poly-t density $p(\delta_i \mid \delta_{\bar{i}})$ are functions (defined in Appendix B.I) of the posterior parameters and of the values of the conditioning variables, while the conditioning values are obtained by the technique used for defining the parameters of STUD. PTDC and PTST use also this technique and the functions that define the parameters of $p(\delta_i \mid \delta_{\bar{i}})$.

III.3 *THE* AI(γ) *APPROACH*

This approach requires the construction of a joint importance function for β and Σ, say $f(\beta, \Sigma)$. Formula (2.19) suggests the following decomposition :

$$(3.26) \qquad f(\beta, \Sigma) = f(\beta \mid \Sigma) \; f(\Sigma)$$

but $f(\beta \mid \Sigma)$ cannot be (2.19) because it is not usually possible to generate random drawings from (2.19). Some suggestions are :

- to take $f(\beta \mid \Sigma)$ as the normal part of (2.19), i.e. $f_N^{\ell_\beta}(\beta \mid \beta_*, V_{\beta\beta})$; see also (2.20);
- to use a rejection procedure based on drawings from that normal density;
- to build a polynomial approximation of $\|B\|^{\tau_*}$ such that random samples can be drawn from the product of that approximation with the normal kernel (e.g. a

quadratic form in β or some elements of it); the possibility to build such an approximation depends on the form of $|B|$.

Defining an importance function $f(\Sigma)$ is also ot easy, because we cannot isolate from (2.16) the kernel of the marginal posterior density of Σ, but we can isolate the kernel of the joint density of β and Σ, say

$$\kappa(\beta,\Sigma) = \|B\|^{\tau*} \, |\Sigma|^{-\frac{1}{2}(\nu_*+m+1)} \exp -\frac{1}{2}\left[\operatorname{tr}\Sigma^{-1} S_0 - \delta_*'(\Sigma^{-1} \,\square\, M_*)\delta_* + (\beta-\beta_*)' V_{\beta\beta}^{-1}(\beta-\beta_*)\right] \tag{3.27}$$

which is needed for computing the ratio $\kappa(\beta,\Sigma)/f(\beta,\Sigma)$ of the kernel of p and f.

The only convenient class of densities for $f(\Sigma)$ is the inverted-Wishart one :

$$f(\Sigma) = f_{iW}^{m}(\Sigma \mid \tilde{S}, \nu). \tag{3.28}$$

The generation of random drawings from inverted-Wishart distributions is discussed in Appendix A.

The matrix $\tilde{S}$ can be chosen, on the basis of (2.12) - (2.13), as :

$$\tilde{S} = S_* + \Delta_*' M_* \Delta_* - \Delta_*' M_* \tilde{E}(\Delta) - \tilde{E}(\Delta') M_* \Delta_* + \tilde{E}(\Delta' M_* \Delta), \tag{3.29}$$

where $\tilde{E}$ denotes some guess of the corresponding expected values, based on guesses of the expected value and of the covariance matrix of δ, such as those provided by (3.8); such guesses could also be based on a preliminary run.

The parameter ν can be fixed at ν_*, so that the factor involving $|\Sigma|$ drops out of the ratio κ/f which takes the value

$$\frac{\kappa}{f} = \|B\|^{\tau_*} \, |V_{\beta\beta}|^{\frac{1}{2}} \exp -\frac{1}{2}\left[\operatorname{tr}\Sigma^{-1}(S_0 - S) - \delta_*'(\Sigma^{-1} \,\square\, M_*)\delta_*\right] \tag{3.30}$$

whenever $f_N^{\ell\beta}(\beta \mid \beta_*, V_{\beta\beta})$ is present in the kernel of $f(\beta \mid \Sigma)$.

The importance function can also be decomposed as

$$f(\beta,\Sigma) = f(\Sigma \mid \beta)\; f(\beta) \tag{3.31}$$

where $f(\Sigma \mid \beta)$ is an inverted-Wishart density approximating $p(\Sigma \mid \beta)$, with parameters $\tilde{S}$, given by an adaptation[8] of (3.29), and ν_*, while $f(\beta)$ is a suitable importance function; possible candidates for $f(\beta)$ are the marginal densities of β obtained from the importance functions of δ proposed in section III.2.

On the whole, this approach seems less promising than the $AI(\Sigma)$ one. Firstly, the quality of the importance functions is more doubtful, though this is an empirical question; there may be cases where they would work. Secondly, this approach is less adapted to the computation of expected values of functions of γ and β, and even of γ only. Indeed, from (2.18), it is clear that $E(\gamma|\beta,\Sigma)$ and $V(\gamma|\beta,\Sigma)$ can be integrated analytically with respect to β, but not with respect to Σ. The latter marginalization should be performed by Monte Carlo, using random drawings of Σ, but even this is not feasible, since the kernel of the marginal density of Σ is not known. As a consequence, these moments must be marginalized by Monte Carlo with respect to β and Σ. If, in addition, expected values of functions of β and γ, whose conditional expectations are not known analytically (e.g. reduced form coefficients), must be computed, random drawings must be generated from (2.18) also. We would then be integrating over the space of δ and Σ by Monte Carlo, which is not desirable.

In any case, the ratio of the number of parameters in γ and Σ, $\ell_\gamma / .5m(m+1)$, is usually less than 1, but not much, unless m is large. Hence for small models (m < 5, say), there is no "dimensional" argument tending to privilege one of the two approaches and the quality of the importance functions is the important issue.

In the seemingly unrelated regression model, β does not exist so that δ and γ coincide, and the choice of one of the two approaches is less clear. The posterior density can indeed be factorized into a normal density for δ, conditional on Σ — see (2.17) —

$$(3.32) \quad p(\delta|\Sigma) = f_N^\ell(\delta|\delta_*, [\Sigma^{-1} \square M_*]^{-1})$$

and a marginal density for Σ — see (2.16) —

$$(3.33) \quad p(\Sigma) \propto |\Sigma|^{-\frac{1}{2}(\nu_*+m+1)} |\Sigma^{-1} \square M_*|^{-\frac{1}{2}} \exp -\frac{1}{2}[\text{tr }\Sigma^{-1}S_0 - \delta_*'(\Sigma^{-1} \square M)\delta_*].$$

Given an inverted-Wishart importance function $f(\Sigma)$ and a random sample $\Sigma_1, \ldots, \Sigma_N$ drawn from it, marginal moments of functions of δ can be computed as

$$(3.34) \quad \frac{1}{N} \sum_{k=1}^{N} E[g(\delta)|\Sigma_k] \frac{\kappa(\Sigma_k)}{f(\Sigma_k)}$$

provided the conditional expectation of $g(\delta)$ is known analytically. Such g functions include polynomials in δ and marginal densities of δ. Note that formula (3.34) (with β replacing δ) is not suited to the case (3.26), since the conditional moments of β are not known analytically.

CHAPTER IV : REPORT AND DISCUSSION OF EXPERIMENTS

IV.1. *REPORT*

We have used the importance functions proposed in section III.2 with 5 models covering a variety of situations, as summarized in the following table :

TABLE 1 : Main features of models used

	BBM	Johnston	Klein	EX	W
number of					
... stochastic equations (m)	2	2	3	2	6
... identities ($\tilde{m}-m$)	0	1	4	0	0
... structural coefficients (ℓ)	2 + 2	1 + 2	3 + 3 + 3	1 + 7	6 × 3
... coeff. of endogenous variables (ℓ_β)	2	2	4	1	0
... coeff. of predetermined variables (ℓ_γ)	2	1	5	7	18
... parameters in Σ (ℓ_Σ)	3	3	6	3	21
$\ell_\gamma / \ell_\Sigma$	.66	.33	.83	2.33	.86
determinant of B matrix of (1.1)	$\neq 1$	$\neq 1$	$\neq 1$	$\equiv 1$	$= 1$
number of observations (T)	16	10	21	11	22
prior	informative	diffuse	diffuse	informative	diffuse
references of previous Bayesian studies	Morales (1971) Richard (1973)	KVD (1978) VDK (1981, 1982 b)	VDK (1980)	Bauwens & d'Alcantara (1981)	(*)

(*) we thank Jean-Paul Lambert for allowing us to use this model.

N.B.: in all models except EX, there are m additional parameters which are the intercept terms, but we have expressed the data in deviations from their means. Posterior moments of these coefficients can be obtained as follows : denoting by ϕ the $m \times 1$ vector of intercepts, by $\bar{y}$ the $m \times 1$ vector of means of the normalized endogenous variables of the stochastic equations, and by $\bar{x}'$ the $m \times \ell$ matrix

$$\begin{bmatrix} \bar{X}'_{(1)} & 0 & \cdots & 0 \\ 0 & \bar{X}'_{(2)} & \cdots & 0 \\ \vdots & \vdots & \ddots & \vdots \\ 0 & 0 & \cdots & \bar{X}'_{(m)} \end{bmatrix}$$

where $\bar{X}'_{(i)}$ is the $1 \times \ell_i$ vector of means of the columns of $X_{(i)}$, as $\phi = \bar{y} - \bar{x}'\delta$, $E(\phi|X) = \bar{y} - \bar{x}'E(\delta|X)$, $V(\phi|X) = \bar{x}'V(\delta|X)\bar{x}$, $\mathrm{Cov}(\phi,\delta|X) = -\bar{x}'V(\delta|X)$.

Similarly, if π_0 is the $m \times 1$ vector of intercepts of the reduced form, $\bar{\tilde{y}}$ the $\tilde{m} \times 1$ vector of means of the columns of $\tilde{Y}$, vec Π the column expansion of $\Pi(mn \times 1)$, and $\bar{Z}'$ the $m \times mn$ matrix

$$\begin{bmatrix} \bar{Z}' & 0 & \cdots & 0 \\ 0 & \bar{Z}' & & 0 \\ \vdots & & \ddots & \vdots \\ 0 & & & \bar{Z}' \end{bmatrix}$$

where $\bar{Z}$ is the $1 \times n$ vector of means of the columns of Z, as $\pi_0 = \bar{\tilde{y}} - \bar{Z}' \text{vec}\,\Pi$, $E(\pi_0 | X) = \bar{\tilde{y}} - \bar{Z}' E(\text{vec}\,\Pi | X)$, $V(\pi_0 | X) = \bar{Z}' V(\text{vec}\,\Pi | X) \bar{Z}$ and $\text{cov}(\pi_0, \text{vec}\,\Pi | X) = -\bar{Z}' V(\text{vec}\,\Pi | X)$.

For each model, we present the equations, the prior density, the importance functions, and some posterior results. We want to emphasize that our purpose is mainly to illustrate how the importance functions behave; hence we do not concentrate much on the interpretation of the posterior results, on issues of specification, or on the elicitation of a prior density.

IV.1.1. BBM

The model of the Belgian beef market used by Morales (1971) and Richard (1973) can be formulated as

$$y_1 = \beta_1 y_2 + \gamma_4 z_4 + u_1 \quad \text{(supply)}$$
$$y_1 = \beta_2 y_2 + \gamma_3 z_3 + u_2 \quad \text{(demand)}$$

where y_1 is the quantity consumed per capita
y_2 is the price index
z_3 is the national income per capita
z_4 is the cattle stock.

The data can be found in Richard (1973, p. 201)[9]

We use the prior A4' of Richard (1973, p. 170 and 184); it is an extended natural-conjugate (ENC) prior on $\delta = (\beta_1 \gamma_4 \beta_2 \gamma_3)'$ and Σ, which is in fact noninformative on β_1 and β_2. The marginal prior densities of γ_4 and γ_3 are independent univariate Student - see (2.4) - with 6 degrees of freedom and

$$E(\gamma_4) = 1.10 \qquad V(\gamma_4) = (.1245)^2$$
$$E(\gamma_3) = .415 \qquad V(\gamma_3) = (.1066)^2,$$

which corresponds to the information built by Morales (1971) from cross section data.

Note that the marginal prior expectation of Σ does not exist, by proposition 2 of section II.2, since the prior moments of β_1 and β_2 do not exist.

Table 2 reports the main characteristics of the importance functions and of the posterior density. Figures 1 to 4, regrouped in Appendix C, show the marginal importance functions and posterior density of each structural coefficient, and the corresponding posterior distribution function.

Let us make a few comments about these results :

1) Comparison of the moments of order 1 and 2 of each importance function with those of the posterior reveal clearly that all the importance functions satisfy rather well the criteria of "location" and "variance-covariance structure" stated in section III.1. In particular, even PTFC essentially satisfies the second criterion, since the posterior correlations[10] between the coefficients of the 2 equations are all rather small.

2) However, the graphs of figures 1 to 4 show that the marginal importance functions based on 1-1 poly-t densities are better approximations of the marginal posterior densities than those based on the Student density. Consider, for example, PTST-1, which factorizes into the product of the conditional poly-t $p(\beta_1, \gamma_4 \mid \beta_2, \gamma_3)$ and of the marginal Student $f(\beta_2, \gamma_3)$; from figures 1-B to 4-B, it can be seen that the marginal importance functions of β_1 and γ_4 are closer to the marginal posterior than those of β_2 and γ_3. On the contrary, for PTST-2, the marginal importance functions of β_2 and γ_3 are better approximations of the posterior densities than those of β_1 and γ_4.

3) The gain in precision obtained by the 2 round procedures applied to STUD, PTST-1 and PTST-2 (see the comments following Table 2) seems worthwhile as is shown by the decrease of the value of the estimated coefficient of variation of the reciprocal of the integrating constant of the posterior density.

	ROUND 1	ROUND 2
STUD	1.97	1.15
PTST-1	1.01	.59
PTST-2	1.63	.73

This is not the case with PTFC and PTDC.

4) The posterior results reported in Table 2 differ slightly from those reported by Richard (1973, p. 205). The relative errors (in %) for the posterior expectations (μ) and standard deviations (σ) of the coefficients, of our results with respect to those of Richard, are

	β_1	γ_4	β_2	γ_3
μ :	- 4.2	.0	5.2	- 1.8
σ :	.0	.0	- 11.5	- 11.1

They indicate that our results underestimate Richards's ones, in absolute value. But there are a few differences between the two approaches, so that the two sets of results are not completely comparable. Let us make a few remarks about this :

i) Richard uses 3 dimensional Gaussian integration, on β_1, β_2 and ρ , the correlation coefficient between the structural disturbances; his ranges of integration are : $-.2 < \beta_1 < 1.2$, $-4 < \beta_2 < 0$, $-1 < \rho < 1$. As ρ is not truncated, it has in principle the same effect in the two methods. In order to assess the influence of the truncation of the ranges of β_1 and β_2 one can simply reject drawings for which β_1 or β_2 do not fall in the above ranges. However, only 44 drawings (out of 10044) are rejected, and the results of Table 2 are not changed, which means that these drawings have negligible weights (κ/f ratio). This is also true of many accepted drawings for which β_1 and β_2 are not "close" to their expected values. We have checked that in Gaussian integration, such points are given more weight, relatively, than in Monte Carlo integration[11]; this could explain our underestimation of some standard deviations.

ii) Out of the 44 rejected drawings, there are 26 for which β_2 falls out of his range; out of these 26, there are 24 for which $\beta_2 > 0$. Again, as Gaussian integration gives more weight than Monte Carlo integration to extreme points, the truncation of the range of β_2 at -4 and 0 in the former approach should result in a smaller expected value of β_2, as observed (- .77 versus - .73). A similar conjecture holds for β_1, since there are 36 rejected drawings for which $\beta_1 < -.2$, but only 4 for which $\beta_1 > 1.2$, so that it is not surprising that Gaussian integration yields a larger expected value of β_1 than Monte Carlo (.25 versus .24).

iii) The underestimation of the standard deviations of β_2 and γ_3 is not due to a lack of convergence of the results by Monte Carlo, as we have checked by looking at the results obtained with less than 10000 drawings. Convergence is achieved even after 1000 drawings.

It seems therefore that the results obtained by Monte Carlo are reliable.

5) The expected values of the covariance matrix Σ and Ω of the disturbances of the structural and reduced forms are

$$E(\Sigma|X) = \begin{bmatrix} .16\,E-1 & \\ -.43\,E-2 & .44\,E-1 \end{bmatrix}$$

$$E(\Omega|X) = \begin{bmatrix} .95\,E-2 & \\ .33\,E-2 & .73\,E-1 \end{bmatrix}$$

Note that $E(\Sigma|X)$ is computed analytically as indicated in formulas (2.12) and (2.13), while $E(\Omega|X)$ is computed by Monte Carlo as described in section II.5 before formula (2.24). These results are also consistent with those of Richard.

TABLE 2 : Results for BBM (N = 10000, + 1000 *for round* 1)

N.B. For a detailed description of the contents of this table, see the comments under the title "Interpretation of Table 2" on the next page.

I.F.	MOMENTS					CORRELATION MATRICES				
	β_1	γ_4	β_2	γ_3		β_1	γ_4	β_2	γ_3	
STUD	.23	1.10	-.71	.56	: μ	1.				β_1
ROUND 2	.09	.07	.19	.07	: σ	-.16	1.			γ_4
95 sec.	–	–	–	–	: γ_1	-.07	.01	1.		β_2
% 4.5	3.1	.4	2.7	1.2	: ε	.02	-.02	-.57	1.	γ_3
PTFC	.24	1.09	-.76	.57	: μ	1.				β_1
	.10	.07	.25	.08	: σ	-.18	1.			γ_4
144 sec.	.30	-.09	-1.73	1.15	: γ_1	.0	.0			β_2
% 1.4	2.4	.3	1.3	.6	: ε	.0	.0	-.72	1.	γ_3
PTDC	.25	1.09	-.76	.57	: μ	1.				β_1
	.12	.07	.26	.08	: σ	-.24	1.			γ_4
2078 sec.	1.36	-.42	-1.78	1.13	: γ_1	-.03	.00	1.		β_2
% 1.5	2.1	.3	1.4	.5	: ε	.02	-.01	-.70	1.	γ_3
PTST-1	.25	1.09	-.75	.56	: μ	1.				β_1
ROUND 2	.13	.07	.22	.07	: σ	-.20	1.			γ_4
972 sec.	-.29	-.09	–	–	: γ_1	-.04	-.05	1.		β_2
% 2.	2.0	.3	1.7	.8	: ε	-.09	.02	-.63	1.	γ_3
PTST-2	.25	1.09	-.77	.57	: μ	1.				β_1
ROUND 2	.12	.07	.29	.09	: σ	-.12	1.			γ_4
939 sec.	–	–	-1.21	.80	: γ_1	-.05	-.02	1.		β_2
% 2.3	1.3	.3	2.0	.6	: ε	-.11	.01	-.67	1.	γ_3
POSTERIOR	.24	1.09	-.73	.56	: μ	1.				β_1
(computed	.11	.07	.23	.08	: σ	-.13	1.			γ_4
using PTFC)	.86	-.08	-1.4	.90	: γ_1	.00	-.10	1.		β_2
	.21	1.09	-.61	.54	: mode	-.17	.03	-.65	1.	γ_4
	.08	.06	.15	.06	: σ_{mode}					

μ = expected value; σ = standard deviation; γ_1 = skewness coefficient;

ε = estimated relative error bound of posterior expected value;

– indicates a quantity that does not exist.

Interpretation of Table 2

Table 2 contains the main characteristics of the importance functions and of the posterior density. One finds in the title the number of drawings used with each importance function.

Column 1 indicates the nature of the importance functions, according to the nomenclature of section III.2, i.e.

- STUD : see formula (3.8)
- PTFC : see formula (3.9)
- PTDC : see formulas (3.11) and (3.13)
- PTST-i : see formulas (3.11) and (3.18); i is the sequence number of the equation corresponding to δ_i ; e.g. for BBM, $\delta_1 = (\beta_1 \gamma_4)' = \delta_{\bar{2}}$, and $\delta_2 = (\beta_2 \gamma_3)' = \delta_{\bar{1}}$.

For each importance function, one finds a block of 4 lines, divided into three parts :

- the right part (under the title CORRELATION MATRICES) reports the correlation matrix of the coefficients; e.g. for STUD, the simple correlation coefficient between β_2 and γ_3 is -.57 ;
- the central part (under the title MOMENTS) contains
  - the expected values of the coefficients (line 1, symbol μ);
  - their standard deviations (line 2, symbol σ);
  - the skewness coefficients γ_1 (line 3, symbol γ_1) defined as μ_3/σ^3, where μ_3 is the third central moment;
  - the estimated relative error bounds of the estimates of the expected values of the *posterior density* obtained with the corresponding importance function; see formula (2.33) where σ/μ_g is replaced by its estimate, as explained in section II.5 after formula (2.34), and $\alpha = .05$ (line 4, symbol ε).
- the left part (i.e. column 1) reports the following information, under the name of the importance function :

  line 2 : whether the parameters of the importance function have been defined by using the estimated posterior moments of a first round (see below); if it is the case, one finds "ROUND 2" in this line;

  line 3 : the total CPU time used for getting the results; if 2 rounds have been done, it is the total CPU time of the two rounds;

  line 4 : the estimated relative error bound of the estimate of the reciprocal of the integrating constant of the posterior density, obtained with the importance function; see formula (2.33) where $\alpha = .05$ and σ/μ_g is replaced by its estimate s_w/m_w (s_w and m_w are the sample standard deviation and mean, respectively, of the weight function (2.28)).

The last block of Table 2 reports the posterior results computed by using the importance function for which the estimated relative error bound of the estimate of the reciprocal of the integrating constant of the posterior is the lowest. The posterior expected values computed with the different importance functions reported in the table are identical for a two-digit accuracy. These posterior results are presented in the same way as for the importance functions, except that the last 2 lines of this block report the modal values of the posterior density (mode line) and the standard deviation of an estimated asymptotic covariance matrix of the posterior (σ_{mode} line); this covariance matrix is equal to minus the inverse of the Hessian matrix of the log posterior, computed at the posterior mode; as explained in section III.2.2., it is also the covariance matrix of the STUD importance function (3.8) of the first round.

As mentioned above, this importance function was used in a first round of 1000 drawings to get estimates of the posterior moments; these moments served then to define the parameters of the STUD importance function for a second round and are reported in the STUD block of Table 2. The STUD importance functions of the two rounds have 3 degrees of freedom, since this is the value of λ proposed in section III.2.2.

In the case of PTST-i, the Student importance function (3.18) of round 1 is taken as the marginal of (3.8); for round 2, its parameters are defined as follows :

- $\bar{d}_i$ is the posterior expectation of δ_i estimated in round 1;
- $\bar{P}_i$ is such that $\bar{P}_i^{-1}/(\bar{\lambda}_i - 2)$ is equal to the posterior marginal covariance matrix, estimated in round 1;
- $\bar{\lambda}_i = 3$ (as in round 1).

IV.1.2. *Johnston*

This small macro model has been used by KVD (1978) and VDK (1981, 1982 b) to illustrate for the first time how Monte Carlo integration could be used for obtaining Bayesian estimates in econometrics and to assess the performance of several importance functions. The model is formulated as follows :

$$c = \beta_1 y + u_1$$

$$i = \beta_2 y + \gamma_2 i_{-1} + u_2$$

$$y = c + i + z$$

where c is consumer expenditure
y is total expenditure
i is investment
z is exogenous expenditure.

The data can be found in Johnston (1963, Table 9.2).

We use the diffuse prior (2.6), i.e.

$$p(\beta_1, \beta_2, \gamma_2, \Sigma) \propto |\Sigma|^{-\frac{1}{2}(\nu_0+3)}$$

and set $\nu_0 = 6$, so that $\lambda = \nu_* - m + 1 - \mu - \tau_* = 16 - 2 + 1 - 2 - 10 = 3$ and it is meaningful to estimate moments of order 1 and 2 of β_1, β_2 and γ_2. Note that the choices of ν_0 recommended by Zellner ($\nu_0 = 0$), Drèze ($\nu_0 = 3$) or Malinvaud ($\nu_0 = 1$) may not guarantee the integrability of the posterior density on $\mathbb{R}^3$, for none of them results in $\lambda > 0$.

This is of course a crude way to specify a prior; for if the prior ensures the existence of moments of order 1 and 2 of the posterior, it is rather logical to require that the corresponding prior moments also exist. If our purpose was not mainly to illustrate how the importance functions behave, we would probably specify a rather noninformative extended natural-conjugate prior. Anyway, a diffuse prior is probably a good test case for our purpose.

An alternative approach is to use diffuse prior densities on truncated ranges, i.e. uniform densities, for the coefficients; then, prior and posterior moments exist. However, even in this case, some value of ν_0 must be selected, but a conservative choice is allowed (e.g. $\nu_0 = 0$); compare DR (end of section 4). Moreover, there exist almost certainly values of the points of truncation for which the posterior results will be identical under the two approaches (with the same ν_0).

Clearly, each approach has some degree of arbitrariness : the value of ν_0, or the ranges of the parameters. In any case, it would be wise to check the sensitivity of the posterior results with respect to these values, if the model were to be used for decision purposes.

Table 4 contains the results, in the same presentation as Table 2. Figures 5 to 7 (see Appendix C) show the marginal importance functions and posterior densities of β_1, β_2 and γ_2, and the corresponding posterior distribution functions. One finds also in figures 8 and 9 the posterior densities and distribution functions of the short and long-run multipliers (srm and lrm) of exogenous expenditure (z) with respect to total expenditure (y). These multipliers are the following functions of the parameters - compare with KVD (1980, section 4) :

$$\text{srm} = 1/(1 - \beta_1 - \beta_2)$$

$$\text{lrm} = 1/[1 - \beta_1 - \beta_2/(1 - \gamma_2)]$$

and have the posterior moments indicated in Table 3 (prior 3).

We obtain larger moments than KVD (see Table 3, prior 1), especially for the standard deviation of the long-run multiplier, since these moments have been computed without imposing lower and upper bounds to the values of the multipliers; compare again with KVD (1980, sections 4 and 5).

The posterior expectations and standard deviations of β_1, β_2 and γ_2 reported in Table 3 can also be compared with those of KVD (1980, Table 1) and VDK (1981, Table 1). In the first paper, these authors use $\nu_0 = 0$ and, in one case, independent normal densities for β_1, β_2 and $\beta_2/(1-\gamma_2)$ (the long run propensity to spend on investment goods), such that these parameters have a prior probability of .95 to lie in the following intervals : $.2 < \beta_1 < .8$, $0 < \gamma_2 < .8$, $.05 < \beta_2/(1-\gamma_2) < .25$. In the second paper, they use $\nu_0 = 1$ $(= m - 1)$ and independent uniform prior densities on $(-2, +2)$ for β_1, β_2 and γ_2. We summarize their results and ours in the following table.

TABLE 3 : Johnston's model : Posterior expectations and standard deviations for 3 prior densities

Prior	β_1	β_2	γ_2	srm	lrm
1.- KVD (1980) $\nu_0 = 0$ and normal	.43 (.07)	.08 (.03)	.40 (.11)	2.12 (.41)	2.44 (.64)
2.- VDK (1981) $\nu_0 = 1$ and uniform on (- 2, 2)	- .60 (.78)	- .31 (.33)	.30 (.14)	not reported	
3.- Present study $\nu_0 = 6$ and diffuse prior	.50 (.07)	.10 (.03)	.37 (.10)	2.63 (.57)	3.28 (6.50)

The results with prior 1 are rather similar to those with prior 3; using $\nu_0 = 6$ and a diffuse prior has roughly the same effect as using $\nu_0 = 0$ and rather sharp normal prior densities on the parameters, especially on the posterior standard deviations. Compared to the results with prior 2, they illustrate clearly the influence of the normal prior or of the parameter ν_0.

We make a few remarks about the importance functions :

1) The skewness coefficients of the importance functions do not exist (since the moments of order 3 do not exist); we report the skewness coefficients of the marginal posterior densities (although the choice of ν_0 does not guarantee their existence) because they seem to converge and their values estimated with STUD, PTST-1 and PTST-2 are fairly equal, especially in the case of β_1; it can also be seen from the plots of the posterior marginal densities in figures 5 to 7 that the signs of the skewness coefficients are consistent with the shape of the densities.

2) There is a problem of convergence of the results for PTFC and PTDC, in the sense that the estimated coefficient of variation of the reciprocal of the integrating constant of the posterior increases with N, the number of drawings, so that the value of ε is much higher with these importance functions. This is due to their incorrect correlation structure. However, the posterior expected values estimated with PTFC and PTDC are still fairly accurate; this is less true for the posterior standard deviations and correlation coefficients. The following table indicates the relative errors (in %) of the posterior estimates obtained with PTFC and PTDC, with respect to those reported in Table 4 :

		β_1	β_2	γ_2
PTFC :	μ	2.	10.	0.
	σ	-28.5	-33.	10.
PTDC :	μ	-2.	0.	2.7
	σ	28.5	0.	0.

It can be seen that these relative errors of the expected values (μ lines) are often much lower than the corresponding values given in Table 4 (ε lines of PTFC and PTDC).

3) The large computer time for PTDC is due to the following reason : the computation of the value of the importance function by formulas (B.12) has to be done with $L = 1000$ to be sufficiently accurate, whereas $L = 100$ was sufficient in the case of BBM.

TABLE 4 : Results for Johnston's model (N = 10000, + 1000 *for round 1)*

N.B. See the comments following table 2 for a detailed description of the contents of this table.

I.F.	MOMENTS β_1	β_2	γ_2		CORRELATION MATRICES β_1	β_2	γ_2	
STUD	.50	.10	.37	: μ	1.			β_1
ROUND 2	.07	.03	.10	: σ	.74	1.		β_2
112 sec.	–	–	–	: γ_1	.03	–.50	1.	γ_2
% 3.4	1.1	2.0	1.6	: ε				
PTFC	.52	.11	.38	: μ	1.			β_1
	.05	.02	.13	: σ	0.	1.		β_2
140 sec.	–	–	–	: γ_1	0.	–.87	1.	γ_2
% 28.6	4.2	10.4	6.9	: ε				
PTDC	.52	.11	.38	: μ	1.			β_1
	.09	.03	.11	: σ	.23	1.		β_2
18867 sec.	–	–	–	: γ_1	–.00	–.60	1.	γ_2
% 17.4	10.0	19.5	3.7	: ε				
PTST-1	.50	.10	.37	: μ	1.			β_1
ROUND 2	.08	.03	.10	: σ	.66	1.		β_2
999 sec.	–	–	–	: γ_1	.05	–.39	1.	γ_2
% 2.7	.8	1.6	1.6	: ε				
PTST-2	.50	.10	.37	: μ	1.			β_1
ROUND 2	.07	.03	.14	: σ	.67	1.		β_2
1073 sec.	–	–	–	: γ_1	.09	–.55	1.	γ_2
% 1.7	.7	1.3	1.1	: ε				
POSTERIOR	.50	.10	.37	: μ	1.			β_1
(computed	.07	.03	.10	: σ	.80	1.		β_2
using STUD,	-1.49	–.91	.17	: γ_1	.10	–.35	1.	γ_2
ROUND 2)	.53	.11	.36	: mode				
	.05	.02	.09	: σ_{mode}				

μ = expected value; σ = standard deviation; γ_1 = skewness coefficient;
ε = estimated relative error bound of posterior expected value ;
– indicates a quantity that does not exist.

IV.1.3. *Klein*

We adopt the formulation of VDK (1980) (except that we change the sign of β_3):

$$c = \alpha_1 p + \alpha_2 p_{-1} + \alpha_3 w + u_1$$

$$i = \beta_1 p + \beta_2 p_{-1} + \beta_3 k_{-1} + u_2$$

$$w_1 = \gamma_1 x + \gamma_2 x_{-1} + \gamma_3 \text{time} + u_3$$

$$x = c + i + g$$

$$p = x - w_1 - t$$

$$k = k_{-1} + i$$

$$w = w_1 + w_2$$

where c is consumer expenditure
i is net investment
w_1 is the wage bill of the private sector
x is net private production
p is profits of the private sector
k is the capital stock
w is the total wage bill
g is the government nonwage expenditure, including the net foreign balance
t is business taxes
w_2 is the government wage bill.

The data can be found in Klein (1950).

We consider essentially a diffuse prior proportional to $|\Sigma|^{-\frac{1}{2}(\nu_0+4)}$ with the following particular cases :

- prior 1 : $\nu_0 = 8$, so that moments of order 1 and 2 of the posterior exist certainly;
- prior 1': $\nu_0 = 8$, and uniform prior densities on the interval (0,1) for all the coefficients except β_3 whose range is (-1, 0);
- prior 2 : $\nu_0 = 6$, so that posterior moments may not exist;
- prior 2': as prior 1', except that $\nu_0 = 6$;
- prior 3': as prior 1', except that $\nu_0 = 0$.

Before discussing the importance functions, we gather the main posterior results in Table 5. Plots of posterior marginal densities for prior 1 and 1' are provided in figures 10 to 18 (Appendix C).

TABLE 5 : Klein's model : Posterior expectations and standard deviations

Prior	α_1	α_2	α_3	β_1	β_2	β_3	γ_1	γ_2	γ_3	importance function used
1	.08 (.13)	.22 (.10)	.79 (.03)	−.19 (.26)	.80 (.19)	−.17 (.04)	.32 (.06)	.23 (.04)	.20 (.04)	PTST-2 ROUND 2
1'	.15 (.07)	.16 (.07)	.79 (.03)	.09 (.07)	.63 (.09)	−.16 (.03)	.35 (.04)	.21 (.04)	.19 (.03)	PTST-2 ROUND 2
2	.04 (.14)	.25 (.11)	.79 (.04)	−.29 (.28)	.87 (.23)	−.18 (.04)	.32 (.06)	.24 (.05)	.20 (.04)	STUD ROUND 2
2'	.14 (.07)	.17 (.06)	.79 (.03)	.08 (.06)	.64 (.08)	−.16 (.02)	.35 (.04)	.22 (.04)	.18 (.03)	STUD ROUND 2
3'	.11 (.08)	.17 (.06)	.80 (.04)	.06 (.06)	.62 (.09)	−.17 (.03)	.35 (.05)	.21 (.05)	.19 (.04)	PTST-2 ROUND 2

The conclusions are clear :

i) truncation of the parameter space reduces substantially the posterior standard deviations of α_1, α_2, β_1 and β_2 and shifts their expected values away from the closest point of truncation, except in the case of α_2 (but this is due to the structure of the posterior correlation matrix, see table 6-b);

ii) the value of ν_0 hardly changes the results when the parameter space is truncated (compare 1' - 2' - 3'); when this is not the case, ν_0 has more influence on the posterior results, especially for α_1, α_2, β_1 and β_2 (compare 1 - 2). Note that we report posterior moments for prior 2, though they may not exist, since the condition of lemma 1 of section III.2.1 is not fulfilled (i.e. $6-3+1-3 = 1 < r = 2$). Computing by Monte Carlo moments that do not exist (theoretically) is always possible, since finite samples are used, but when the sample size increases, the sample moments usually increase. Since this behaviour was not observed for the posterior moments of order 1, 2 and even 3 under prior 2, it can be supposed that these moments exist.

Our results for prior 3' are very close to those of KVD (1980, Table 4, prior 2); their prior 2 is essentially the same as our prior 3'; in addition, they incorporate two restrictions on the parameter space : the determinant of the B matrix of the structural form - which is equal to $1-(\alpha_1+\beta_1)(1-\gamma_1)-\alpha_3\gamma_1$ - must be larger than .01 in absolute value, and the dominant root of the characteristic polynomial of the reduced form must be less than 1 in absolute value (implying stability of the system).

As regards the importance functions, we present some detailed results for the case of prior 1 in Tables 6-a, 6-b and figures 10-A to 18-A (Appendix C), and for the case of prior 1', in Tables 7-a, 7-b and figures 10-B to 18-B. We can make the following remarks about the results :

1) When the prior is not truncated, it turns out that two importance functions (PTST-2 and PTST-3, in a second round) behave quite satisfactorily, in the sense that no problem of convergence of the results seem to occur. PTST-3 is a little less efficient than PTST-2 : e.g. after 10000 drawings, the estimated relative error bound of the reciprocal of the integrating constant of the posterior is 14.3 (%) with PTST-3, but 10.8 % with PTST-2. We indicate briefly what happens with the other importance functions :

   i) STUD (in a second round) and PTDC still provide fairly accurate estimates of the posterior moments, even though the *estimated* variation coefficient of the reciprocal of the integrating constant of the posterior increases with the number (N) of drawings; e.g. with STUD, in a particular run, it takes the values 6.622 (N = 10000), 6.882 (N = 20000) and 16.50 (N = 50000), the latter value corresponding to the estimated relative error of 28.9 % in table 6-a. PTDC requires an excessive computer time because the value of L must be of the order of 10000 to apply formula (B.12) for computing the importance function with sufficient precision.

   ii) PTFC is obviously a poor approximation of the posterior in view of the correlation structure of the latter (see Table 6-b).

   iii) PTST-1 does not work at all : after a first round, it is not even correctly located with respect to the parameters α_1, α_2, β_1 and β_2. As is well known, these are the parameters creating most of the variation of the posterior. In all cases of PTST, the marginal Student importance function $f(\delta_{\bar{1}})$ has been defined as the marginal, in round 1, of (3.8), and in round 2, of the global Student approximation built upon the posterior results of round 1. However, ideally $f(\delta_{\bar{1}})$ should be directly built as an approximation of $p(\delta_{\bar{1}})$ whose kernel is given by (3.4), i.e. by computing the mode of (3.4) and the corresponding covariance matrix. The two round procedure is an indirect way to do this, but the approximation may be too coarse in the case of $f(\delta_{\bar{1}})$.

2) When the prior is truncated, it turns out that STUD is also a good importance function, as PTST-2 and PTST-3. As can be expected, truncation results in smaller estimated coefficients of variation and relative errors. But there is a price to pay, for one has to reject the drawings which are not in the prior ranges of the parameters. For PTST-2, the rate of rejection is about 2/3: for STUD, it is about 1/2. Note that a second round for STUD was not used in this case.

3) The benefit resulting from using PTST-2 when the prior is not truncated is illustrated by the plots of the marginal importance functions and posterior densities (figures 10 to 18 of Appendix C). It can be seen in particular that the marginal importance functions for β_1, β_2 and β_3 corresponding to PTST-2 are better approximations of the posterior than the Student ones, as can be expected from the use of the 1-1 conditional poly-t for $\beta_1, \beta_2, \beta_3$ in PTST-2. This is also the case for α_2 and especially α_1, although these marginal importance functions are Student in both cases. Indeed, as α_1 and α_2 are rather strongly correlated with β_1 and β_2, the parameters of their importance functions are better revised after the first round with PTST-2 than after the first round with STUD. It follows that the second round Student marginal importance function for α_1 and α_2 is a better approximation with PTST-2 than with STUD.

When the prior is truncated, this effect is less marked. In general, STUD and PTST-2 are closer to each other in figures 10-B to 18-B than in figures 10-A to 18-A. A comparison of the results in Tables 6-b and 7-b also reveals that the correlation matrices of STUD and PTST-2 are much more similar to the posterior correlation matrix when the prior is truncated (Table 7-b) than when it is not (Table 6-a).

4) When ν_0 is too low, the conditional 1-1 poly-t (3.3) are not defined since the integral (3.5) does not converge if $2(\lambda_1 - \lambda_0) - \ell_i \leq 0$; this means that ν_0 must be strictly larger than $\frac{1}{2}\ell_i$. With $\nu_0 < \frac{1}{2}\ell_i$ in the prior, we can use PTST-i provided we increase the value of λ_1. Hence the quality of the importance function defined in this way may be doubtful. A Student importance function can also be used. For computing the posterior results with prior 3' (see Table 5), we used PTST-2 (with $\lambda_1 = 12.75$) and STUD, with 1000 accepted drawings. It seems that convergence has been achieved, but given the limited number of drawings, the results are certainly less precise in this case.

5) An additional difficulty in the case of Klein's model, for all poly-t based importance functions, is the odd number of observations; indeed the value of λ_0 in the conditional 1-1 poly-t density (3.3) is equal to $\frac{1}{2}T$ (since $\tau_0 = 0$), but when this value is half integer, it is not possible to generate random drawings from the 1-1 poly-t[12]; this density must be replaced by another one with the same parameters, except that λ_0 is fixed at $\frac{1}{2}(T-1)$. This introduces additional variation in the weight function (κ/f ratio).

TABLE 6-a : *Results for Klein's model (prior 1)* (N = 50000, + 1000 *for round 1)*

N.B.: See the comments after Table 2 for a detailed description of the contents of this table and of Table 6-b.

I.F.	MOMENTS									
	α_1	α_2	α_3	β_1	β_2	β_3	γ_1	γ_2	γ_3	
STUD	.09	.23	.79	−.15	.82	−.18	.33	.22	.19	: μ
ROUND 2	.08	.09	.04	.20	.18	.04	.07	.04	.03	: σ
2947 sec.	–	–	–	–	–	–	–	–	–	: γ_1
% 28.9	91.0	16.2	1.1	57.3	7.9	6.5	4.2	6.2	6.5	: ε
PTST-2	.06	.24	.78	−.28	.87	−.18	.33	.23	.20	: μ
ROUND 2	.13	.10	.03	.32	.23	.04	.05	.03	.04	: σ
8317 sec.	–	–	–	–	–	–	–	–	–	: γ_1
% 5.1	13.3	2.9	.2	8.6	1.4	1.1	1.2	1.2	1.1	: ε
POSTERIOR	.08	.22	.79	−.19	.80	−.17	.32	.23	.20	: μ
(computed	.13	.10	.03	.26	.19	.04	.06	.04	.04	: σ
using PTST-2,	−1.71	.93	.01	−1.13	1.05	−.70	−.58	.40	.43	: γ_1
ROUND 2)	.15	.17	.78	.01	.69	−.17	.35	.22	.18	: mode
	.08	.07	.03	.15	.11	.02	.04	.03	.03	: σ_{mode}

μ = expected value; σ = standard deviation; γ_1 = skewness coefficient ;

ε = estimated relative error bound of posterior expected value.

TABLE 6-b : Results for Klein's model (prior 1) : correlation matrices

I.F.		α_1	α_2	α_3	β_1	β_2	β_3	γ_1	γ_2	γ_3
STUD	α_1	1.								
ROUND 2	α_2	-.74	1.							
	α_3	-.05	-.34	1.						
	β_1	.48	-.53	-.26	1.					
	β_2	-.59	.73	-.14	-.73	1.				
	β_3	.18	-.33	.47	.12	-.53	1.			
	γ_1	-.05	.06	-.64	.61	-.04	-.57	1.		
	γ_2	.05	-.13	.56	-.47	.09	.58	-.88	1.	
	γ_3	.14	.20	.00	-.44	-.00	.38	-.64	.40	1.
PTST-2	α_1	1.								
ROUND 2	α_2	-.84	1.							
	α_3	-.04	-.27	1.						
	β_1	.47	-.52	.12	1.					
	β_2	-.45	.50	-.15	-.91	1.				
	β_3	-.02	-.02	.08	.36	-.48	1.			
	γ_1	-.02	-.17	-.06	.27	-.09	-.22	1.		
	γ_2	.04	-.03	-.02	-.16	.06	.29	-.81	1.	
	γ_3	-.07	.40	.40	-.25	.09	.22	-.77	.5	1.
POSTERIOR	α_1	1.								
	α_2	-.77	1.							
	α_3	-.27	-.10	1.						
	β_1	.60	-.55	-.03	1.					
	β_2	-.49	.63	-.10	-.79	1.				
	β_3	-.12	-.02	.24	.24	-.40	1.			
	γ_1	.11	-.06	-.19	.43	-.08	-.45	1.		
	γ_2	-.09	.03	.12	-.29	.16	.50	-.83	1.	
	γ_3	.04	.16	-.33	-.22	-.09	.38	-.67	.40	1.

TABLE 7-a : Results for Klein's model (prior 1') (N = 50000, + 1000 *for round 1)*

N.B.: See the comments after Table 2 for a detailed description of the contents of this table and of Table 7-b.

I.F. (TRUNCATED)	MOMENTS α_1	α_2	α_3	β_1	β_2	β_3	γ_1	γ_2	γ_3	
STUD	.16	.16	.78	.09	.65	-.17	.37	.21	.18	: μ
	.06	.05	.03	.08	.08	.02	.03	.03	.03	: σ
3725 sec.	.68	.20	-.51	2.42	-1.29	-.04	1.01	-.90	-.01	: γ_1
% 8.3	4.5	3.9	.2	5.3	1.1	1.9	2.1	2.8	2.2	: ε
PTST-2	.17	.15	.78	.09	.64	-.17	.36	.21	.18	: μ
ROUND 2	.06	.05	.03	.07	.08	.02	.04	.03	.03	: σ
17059 sec.	.66	.21	-.82	1.15	-.15	-.30	.72	-.63	.13	: γ_1
% 3.1	1.9	1.6	.2	2.7	.5	.7	.6	.8	.8	: ε
POSTERIOR	.15	.16	.79	.09	.63	.16	.35	.21	.19	: μ
(computed using PTST-2,	.07	.07	.03	.07	.09	.03	.04	.04	.03	: σ
ROUND 2)	.15	.22	-.10	1.01	-.01	-.14	-.60	.41	.50	: γ_1

μ = expected value; σ = standard deviation; γ_1 = skewness coefficient;

ε = estimated relative error bound of posterior expected value.

TABLE 7-b : Results for Klein's model (prior 1') : correlation matrices

I.F. (truncated)		α_1	α_2	α_3	β_1	β_2	β_3	γ_1	γ_2	γ_3
STUD	α_1	1.								
	α_2	-.53	1.							
	α_3	-.35	-.25							
	β_1	.01	-.08	.04	1.					
	β_2	-.14	.33	-.08	-.55	1.				
	β_3	-.02	-.08	.14	.19	-.32	1.			
	γ_1	-.46	.29	-.04	.32	.12	-.50	1.		
	γ_2	.34	-.40	.02	-.25	.05	.52	-.85	1.	
	γ_3	.43	.10	-.51	-.14	-.22	.35	-.61	.34	1.
PTST-2	α_1	1.								
ROUND 2	α_2	-.52	1.							
	α_3	-.42	-.18	1.						
	β_1	.06	-.10	-.00	1.					
	β_2	-.10	.22	-.02	-.61	1.				
	β_3	-.07	.03	.08	.26	-.37	1.			
	γ_1	-.34	.13	.00	.34	.05	-.31	1.		
	γ_2	.25	-.27	-.03	-.27	.09	.34	-.84	1.	
	γ_3	.44	.15	-.55	-.13	-.17	.22	-.61	.33	1.
POSTERIOR	α_1	1.								
	α_2	-.52	1.							
	α_3	-.38	-.20	1.						
	β_1	.02	-.09	.03	1.					
	β_2	-.10	.30	-.08	-.57	1.				
	β_3	-.06	-.03	.14	.22	-.33	1.			
	γ_1	-.43	.23	-.04	.30	.10	-.47	1.		
	γ_2	.33	-.36	-.00	-.23	.07	.49	-.83	1.	
	γ_3	.42	.14	-.50	-.12	-.20	.34	-.60	.33	1.

IV.1.4. EX

This is a 2 equation model of exports and export price of the Belgian industrial sector studied by Bauwens and d'Alcantara (1981) in the Bayesian framework and used by Drèze and Modigliani (1981), hereafter DM, to get estimates of the elasticity of export prices with respect to factor costs and of the elasticity of exports quantities with respect to the same variable. We use the following version of the model[13]:

$$\ell n \frac{px}{pxw} = \alpha \, \ell n \frac{pb}{pxw} + u_1, \quad (0 < \alpha < 1)$$

$$\ell n \, qx = \beta_0 + \beta_1 \, \ell n \, qmw + \beta_2 \, \ell n \frac{px}{pxw} + \beta_3 \, \ell n \left(\frac{px}{pxw}\right)_{-1} + \beta_4 \, \Delta \, \ell n \, qmw + \beta_5 \, \ell n \left(\frac{px}{pb}\right)_{-1} + \beta_6 \, \ell n \, qp_{-1} + u_2$$

where px = index of export prices
pxw = index of competitors price
pb = index of factor costs
qx = export volume
qmw = measure of foreign demand
qp = potential output.

For more derails on these variables and the interpretation of the equations, see Bauwens and d'Alcantara (1981).

These authors have also introduced prior information on several parameters, in the form of an independent beta density on α, on the interval (.05, .95), with parameters 3 and 6, implying a mode of .28, an expectation of .33, and a standard deviation of .135, so that the prior probability that α is less than .5 is rather high (.86). A marginal Student density on $\beta_1, \beta_2, \beta_3, \beta_5, \beta_6$ completes the prior, together with a diffuse prior on Σ, β_0 and β_4 proportional to $|\Sigma|^{-\frac{3}{2}}$.

In the present application, we have used an extended natural-conjugate prior approximating the prior described above. This is done in the following way :

i) we define a Student density on α, centered at .28, with standard deviation equal to .15, and 16 degrees of freedom, and compute the scalar M_{011} of the extended natural-conjugate prior (2.3) as

$$M_{011} = \frac{s_{01}}{16-2} (.15^2)^{-1},$$

where s_{01} is to be chosen below. Next, we truncate this Student density in the interval (0 ,1). This results in a prior skewed to the right as in the case of the beta density described above.

ii) Similarly, we define a Student density on $(\beta_1 \ \ \beta_2 \ \ \beta_3 \ \ \beta_5 \ \ \beta_6)$ centered at (1. −.75 −.75 1.5 1.33); the corresponding standard deviations are (.25 .99 .99 .52 .16); all correlations are 0. except between β_2 and β_3 (−.5). These moments are identical to those elicited by Bauwens and d'Alcantara (1981), who describe in detail how they have been chosen. Then, M_{022} is built in the same way as M_{011}, i.e.

$$M_{022} = \frac{s_{02}}{10-2} V_{022}^{-1}$$

(where V_{022} is the prior covariance matrix described above); see also formula (2.5).

iii) The two Student densities have the same exponent $\frac{1}{2}\nu_0$, where $\nu_0 = 17$, which is also the number of degrees of freedom of the inverted-Wishart prior on Σ. The matrix S_0 of this prior is taken as diag (s_{01}, s_{02}) but one is faced with the delicate problem of fixing s_{01} and s_{02}. Noting that $S_0/\nu_0 - 3$ is the prior expectation of Σ given that $\Delta = \Delta_0$ - see (2.14) - the diagonal elements of the matrix S_0 can be interpreted as the residual sums of squares from a hypothetical sample of size $\nu_0 = 17$. We have fixed them at (roughly) $\nu_0/T = \frac{17}{11}$ times the residual sum of squares of the price equation estimated by FIML, on the observed sample (i.e. $s_{01} = 56$) and 6 times the residual sum of squares of the quantity equation (i.e. $s_{02} = .028$). We use a factor of 6 (instead of $\frac{17}{11}$) for s_{02} because otherwise M_{022} has not enough weight with respect to the sample matrix M_{22}. Given these values of s_{01} and s_{02}, the prior equivalent of the FIML estimated standard errors of the two equations (2. and .02) are 2. and .04. In this way, we try not to contradict too much the sample information on Σ, while at the same time giving significant weight to the prior information on the coefficients (especially β_2, β_3 and β_5). This is shown by a comparison of the matrices M_0 and M of (2.15) :

$$M_0 = \begin{bmatrix} 177.78 & & & & & & & \\ 0 & 0 & & & & & & \\ 0 & 0 & .0583 & & & & & \\ 0 & 0 & 0 & 47.62 & & & & \\ 0 & 0 & 0 & 23.81 & 47.62 & & & \\ 0 & 0 & 0 & 0 & 0 & 0 & & \\ 0 & 0 & 0 & 0 & 0 & 0 & 129.63 & \\ 0 & 0 & 0 & 0 & 0 & 0 & 0 & .1167 \end{bmatrix}$$

$$\begin{matrix} \alpha & \beta_0 & \beta_1 & \beta_2 & \beta_3 & \beta_4 & \beta_5 & \beta_6 \end{matrix}$$

$$M = \begin{bmatrix} 363.4 & & & & & & & \\ 44.6 & 11.0 & & & & & & \\ 107.8 & 23.7 & 52.3 & & & & & \\ 52.8 & 7.6 & 17.3 & 48.42 & & & & \\ 10.1 & 1.6 & 4.4 & 2.77 & 37.55 & & & \\ 34.4 & 10.0 & 21.2 & 10.1 & 3.73 & 12.9 & & \\ -245.8 & -30.0 & -76.1 & -28.0 & 18.8 & -27.1 & 267.3 & \\ 113.7 & 26.5 & 57.6 & 19.2 & 4.43 & 23.7 & -78.0 & 63.99 \end{bmatrix}$$

The prior thus obtained is probably too data-based to be acceptable for solving a real life decision problem, but in this illustration of the performance of importance functions we do not want to have one source of information dominated by the other.

Table 8 contains the posterior results for the coefficients (except β_0) and some elasticities, obtained with 3 prior densities :

- DM : a diffuse prior proportional to $|\Sigma|^{-\frac{3}{2}}$, with the exact prior restriction $\alpha + \beta_6 = 1$ together with $.05 < \alpha < .95$;
- Bd'A : the prior of Bauwens - d'Alcantara described above;
- ENC : the extended natural-conjugate prior described above which may be viewed as an approximation to Bd'A.

Note that for the Bd'A and ENC prior, $\alpha + \beta_6$ has a prior expectation equal to 1.61 (= 1.33 + .28) and a prior standard deviation equal to .22. The corresponding posterior values are 1.29 and .19. Indeed the sample information is not in accordance with the prior expectation of β_6 (= 1.33) - the maximum likelihood estimate of β_6 is -.36 - and shifts the prior expectation of β_6 from 1.61 to 1.29. As explained in Bauwens and d'Alcantara (1981), a prior expectation for β_6 (i.e. the elasticity of exports with respect to potential output) greater than 1 is chosen to reflect the tendency of the small open economy of Belgium to be more and more integrated in international trade; for the indicated prior moments of β_6, the .95 prior probability interval for this elasticity has no values below 1 (using a normal approximation). The unreasonable maximum likelihood estimate of β_6 is due to a strong collinearity in the data, between the variables $\ell n\ qmw$, $\Delta\ \ell n\ qmw$ and $\ell n\ qp_{-1}$; see again Bauwens and d'Alcantara (1981) and footnote 13.

The results for the DM prior were used by DM to estimate short-run and long-run elasticities of employment (L) with respect to real wages (W/P) for Belgium. For example, the short-run elasticity is derived by these authors, taking account that the balance of payments is a binding constraint, as

$$\eta_{L,W/P} = \frac{\eta_{\Xi W} - \eta_{MW}}{\eta_{MY}(1 - \eta_{PW})} \cdot \eta_{LY}$$

where $\eta_{\Xi W}$ is the elasticity of the value of exports ($\Xi = PX \cdot X$) with respect to nominal wages (W)

η_{MW} is the elasticity of import quantities (M) with respect to W

η_{MY} is the elasticity of M with respect to national income (Y)

η_{PW} is the elasticity of the price level (P) with respect to W

η_{LY} is the elasticity of employment (L) with respect to Y, taken to be equal to $\frac{1}{3}$ in the short-run.

Using the EX model, one can compute $\eta_{\Xi W}$ as

$$\eta_{\Xi W} = \alpha + (1 - \eta_{MX})[\alpha(\beta_2 + \beta_3) - (1-\alpha)\beta_5]$$

where η_{MX} is the elasticity of M with respect to X, taken to be equal to .4.

As regards η_{MY} and η_{MW}, DM use a normal approximation of the posterior joint density for these elasticities, obtained from a Bayesian single equation analysis of an import quantity equation, using a noninformative prior. This density has the following mean vector and covariance matrix V :

$$E\left(\begin{bmatrix}\eta_{MY}\\ \eta_{MW}\end{bmatrix}\right) = \begin{bmatrix}1.53\\ .19\end{bmatrix} \qquad V = \begin{bmatrix}.25^2 & \\ .00875 & .18^2\end{bmatrix}$$

Finally, DM use a normal density for η_{PW}, with mean .65, standard deviation .20; this density is truncated at .85.

DM have in fact used a combination of analytical and numerical procedures (essentially the iterative Simpson method of numerical integration) to obtain the posterior density of $\eta_{L,W/P}$. We shall instead apply a Monte Carlo method, which is very easy to implement. Indeed during the integration phase of the posterior density, we use the random drawings of the parameters α, $\beta_0,\ldots,\beta_6$ to compute $\eta_{\Xi W}$; at the same time, we generate random drawings of η_{MY}, η_{MW} and η_{PW} from their normal densities, and compute the corresponding value of $\eta_{L,W/P}$. Finally the posterior expectation of $\eta_{L,W/P}$ is the weighted average of these values, divided by the integral of the posterior kernel (see formula (2.29) and what is said after it).

A similar treatment is applied for computing the posterior moments of the long-run elasticity, whose expression is given by formula (16') of DM.

In Table 8, the results obtained with the prior of Bd'A are very close to those obtained with the ENC prior (except for β_2 and β_3 separately, but still for $\beta_2 + \beta_3$); this indicates that the ENC prior built in this section is , as expected by construction, a fairly good approximation of the Bd'A prior. Although these results for the structural parameters are rather different from those obtained with the DM prior, this is less true for the elasticities. Marginal densities and distribution functions of $\eta_{\Xi W}$, $\eta_{L,W/P}$ (short-run and long-run) are shown in figures 19 to 21 and can be compared with those of DM (see their figures 7, 3 and 5).

Tables 9-a and 9-b report the usual results about the importance functions (we do not report all the correlation matrices since they are very close to the posterior correlation matrix). We make three comments :

1) The B matrix of the model is triangular and its determinant is equal to 1, identically in β_2. Consequently, $p(\delta_i | \delta_{\bar{i}})$, see (3.3), is a Student density and the posterior density of δ can be expected to be almost symmetrical. This is the case, as shown by the values of the skewness coefficients of the posterior (see last line of Table 9-a). The posterior density of α is slightly skew because it is truncated, but this has no effect on the other coefficients because α is almost uncorrelated with them (see last line of Table 9-b).

2) It is interesting to notice that PTST-2 is a much better importance function than PTST-1 (compare the ε values for the reciprocal of the integrating constant of the posterior); indeed this is due to the difference in the number of conditioning variables of $p(\delta_i | \delta_{\bar{i}})$ (1 for i = 2, 7 for i = 1). In both cases, $f(\delta_{\bar{i}})$ is the marginal of STUD and is a better approximation of $\delta_{\bar{1}}$ (= α) than for $\delta_{\bar{2}}$ (= β_i's). It is therefore not surprising that PTST-1 and STUD are very similar.

3) PTFC is the best importance function, which is coherent with the fact that the correlations between α and β's are almost zero.

TABLE 8 : Posterior results for EX : expected values and standard deviations

prior :	DM	Bd'A	ENC
β_1	.62 (.12)	.49 (.10)	.50 (.11)
β_2	-4.13 (6.54)	-.77 (1.09)	-1.14 (.77)
β_3	-2.18 (1.18)	-1.88 (.78)	-1.44 (.75)
β_4	.73 (.31)	.91 (.32)	.81 (.34)
β_5	1.07 (.84)	1.40 (.45)	1.37 (.45)
β_6	.83 (.10)	1.08 (.18)	1.10 (.17)
α	.17 (.10)	.21 (.07)	.19 (.08)
$\alpha + \beta_6$	1. (.0)	1.29 (.19)	1.29 (.19)
$\eta_{\Xi W}$	-.31 (.11)	(*)	-.26 (.09)
$\eta_{L,W/P}$ (short-run)	-.58 (.44)	(*)	-.16 (.21)
$\eta_{L,W/P}$ (long-run)	-1.82 (.79)	(*)	-1.67 (.96)

(*) not available

TABLE 9-a : Results for EX model (N = 10000)

N.B. : See the comments after Table 2 for a detailed description of the contents of this table.

I.F.	MOMENTS β_1	β_2	β_3	β_4	β_5	β_6	α		NR(**)
STUD	.50	−1.14	−1.43	.80	1.36	1.09	.19	: μ	152
137 sec.	.09	.78	.61	.46	.53	.12	.07	: σ	
% 10.1	2.3	11.2	6.3	4.4	3.5	1.8	4.9	: ε	
PTFC	.50	−1.14	−1.44	.81	1.38	1.10	.19	: μ	312
150 sec.	.11	.80	.77	.35	.46	.17	.09	: σ	
% .8	.8	2.7	2.0	1.6	1.3	.6	1.6	: ε	
PTDC	.49	−1.11	−1.39	.79	1.34	1.06	.19	: μ	564
2066 sec.	.14	.82	.80	.38	.52	.26	.10	: σ	
% 1.1	.8	2.7	2.1	1.7	1.3	.6	1.6	: ε	
PTST-1	.50	−1.14	−1.42	.81	1.38	1.09	.19	: μ	339
289 sec.	.09	.80	.61	.52	.58	.13	.09	: σ	
% 9.2	2.2	7.4	5.4	4.0	3.2	1.7	4.1	: ε	
PTST-2	.50	−1.15	−1.42	.81	1.38	1.09	.19	: μ	142
407 sec.	.11	.79	.77	.35	.47	.17	.07	: σ	
% 2.0	.8	2.9	2.2	1.8	1.4	.7	2.3	: ε	
POSTERIOR	.50	−1.14	−1.44	.81	1.37	1.10	.19	: μ	
(computed	.11	.77	.75	.34	.45	.17	.08	: σ	
using PTFC)	−.01	.01	.00	−.01	−.04	.04	.22	: γ_1	

μ = expected value; σ = standard deviation; γ_1 = skewness coefficient ;
ε = estimated relative error bound of posterior expected value.

(**) number of rejected drawings ($\alpha < 0$ or $\alpha > 1$) to get 10000 accepted drawings drawings.

TABLE 9-b : Posterior results for EX : correlation matrix

	β_1	β_2	β_3	β_4	β_5	β_6	α
β_1	1.						
β_2	.05	1.					
β_3	-.15	-.30	1.				
β_4	-.05	-.08	-.15	1.			
β_5	.38	-.09	-.27	.13	1.		
β_6	-.84	-.11	-.00	.23	.06	1.	
α	-.01	-.05	.01	-.02	.01	-.02	1.

IV.1.5. *W*

We consider a model of seemingly unrelated wage equations of 6 countries of the EEC (Germany, France, Italy, Netherlands, Belgium and United Kingdom). The specification is similar for each country

$$d\ell w = \gamma_1 \, d\ell pc + \gamma_2 \, d\ell pr + \gamma_3 \, u + \varepsilon$$

where $d\ell w$ is the growth rate of the real wage rate
$d\ell pc$ is the growth rate of the price index of consumption
$d\ell pr$ is a measure of the growth rate of productivity
u is the rate of unemployment (in some cases, lagged one period).

The variables are expressed in deviations to their means. The data are taken from the data bank of the COMET model; see Barten et al. (1976).

The prior is completely diffuse, being proportional to $|\Sigma|^{-\frac{1}{2}(m+1)}$, where m is the number of equations.

In this type of model, $|B| = 1$ since B is an identity matrix of order m, so that the posterior density of δ may not be too asymmetrical. Consequently, the STUD importance function might be performant, even though the number of parameters is large (18). Usually, the correlation coefficients between the parameters of the different equations are not too large in absolute value, so that PTFC might also be a good importance function.

This is the case with this model. Table 10 gives the results for the 6 countries, using the STUD importance function. For PTFC, it gives the results for 4 countries (Germany, France, Netherlands, Belgium) because the number of observations (22) is not large enough for the S_* matrix - see (2.15) - of the posterior to be positive definite[14]; by lemma 2 of section III.2.1, S_* has to be positive definite in order to get the conditional densities $p(\delta_i \mid \delta_{\bar{i}})$.

We do not report the skewness coefficients of the posterior, because their estimates obtained with STUD and PTFC are too different; this means that they are not yet very precisely estimated with a sample size of 10000. However, in both cases, they are close to zero for all coefficients. An alternative measure of skewness is given by

$$Sk = \frac{\mu - \text{mode}}{\sigma}$$

see e.g. Kendall and Stuart (1963, p. 85).

In Table 10, the μ column of the STUD importance function gives the modal values of the posterior density and it can be easily seen that the posterior expected values are almost equal to the corresponding modal values, especially in units of σ.

This example is also interesting for three reasons :

1) It shows that even in high dimension, it is possible to achieve variance reduction by "sophisticated methods"; this tempers the skepticism of Davis and Rabinowitz (1975, section 5.10) about the possibility to use a method like importance sampling for computing integrals of dimension greater than about 12.

2) The Bayesian analysis of seemingly unrelated regressions is shown to be operational even in high dimension.

3) The numerical elicitation of an extended natural-conjugate prior (2.3) for the simultaneous equation model case is quite feasible provided $\tau_0 = 0$ (a choice we have advocated strongly in section II.2) and of course $\nu_0 > m-1$. In particular there is no need to restrict S_0 to be diagonal, M_0 to be block diagonal, and Δ_0 to be "column-diagonal", as Δ, since all these characteristics are not verified in the posterior density of the seemingly unrelated regression model. Therefore it would be possible to introduce probabilistic dependence between coefficients of different equations (e.g. through M_{0ij}) in the prior. This is a topic for further research.

TABLE 10 : Results for the W model (N = 10000)

N.B. : see the comments following Table 2 for a detailed description of the contents of this table.

		6 equations					4 equations							
		STUD			POSTERIOR		STUD			PTFC			POSTERIOR (computed using STUD)	
		μ	σ	ε	μ	σ	μ	σ	ε	μ	σ	ε	μ	σ
DB	γ_1	.05	.18	38.6	.06	.22	.08	.18	18.1	.08	.21	19.	.08	.21
	γ_2	.65	.19	3.1	.65	.22	.55	.19	2.4	.55	.23	2.2	.55	.21
	γ_3	-.50	.33	6.4	-.54	.38	-.61	.32	4.0	-.61	.37	4.7	-.64	.38
FR	γ_1	.04	.11	31.4	.03	.13	.00	.12	(*)	.00	.14	(*)	.00	.14
	γ_2	.85	.38	5.4	.80	.43	.75	.38	4.4	.74	.46	3.8	.72	.44
	γ_3	.89	.48	5.5	.89	.55	.85	.48	4.2	.85	.53	4.3	.83	.53
IT	γ_1	-.10	.10	11.5	-.10	.11								
	γ_2	.57	.24	4.3	.57	.27								
	γ_3	-.26	.18	6.1	-.27	.21								
NL	γ_1	.15	.09	8.6	.16	.11	.17	.09	3.8	.17	.10	4.1	.17	.10
	γ_2	.78	.12	1.6	.77	.13	.71	.11	1.2	.71	.14	1.1	.72	.13
	γ_3	-1.28	.27	2.4	-1.30	.32	-1.35	.27	1.5	-1.34	.31	1.5	-1.34	.31
BE	γ_1	.29	.11	8.0	.26	.14	.29	.11	3.4	.29	.12	3.7	.28	.13
	γ_2	.52	.20	7.1	.53	.24	.62	.20	2.3	.62	.23	2.4	.62	.22
	γ_3	-.55	.35	8.8	-.55	.44	-.62	.38	6.5	-.62	.43	5.1	-.60	.44
UK	γ_1	.14	.13	13.9	.12	.16								
	γ_2	.12	.24	26.9	.11	.29								
	γ_3	-1.04	.74	8.1	-.93	.85								
ε, integral		8.2 %					4.4 %			5.0 %				
CPU		951 sec.					380 sec.			390 sec.				

(*) not available since the expected value is equal to 0

μ = expected value; σ = standard deviation; ε = estimated relative error bound of posterior expected value.

IV.2. CONCLUSIONS

1) We have shown with different models that some importance functions proposed in section III.2 are quite efficient for computing posterior moments. However they are not all equally efficient, as shown in Table 11, which gives the efficiency ratios of the poly-t based importance functions PTFC, PTDC and PTST-i with respect to STUD. The efficiency ratio of method 2 with respect to method 1 is $t_1\sigma_1^2 / t_2\sigma_2^2$, where t_i is the computer time and σ_i^2 is the estimated variance of the weight function (2.28) of method i; see Hammersley and Handscomb (1979, section 5.1). If the number of drawings used with the two methods is equal, this ratio is equal to $t_1\varepsilon_1/t_2\varepsilon_2$ where ε_i is the estimated relative error bound of the integral to be computed.

TABLE 11 : Efficiency ratios of PTFC, PTDC and PTST-i with respect to STUD

	PTFC	PTDC	PTST-i (*)
BBM	2.12	.14	.22
Johnston	.10	.00	.21
Klein	-	.00	2.01
Klein (truncated prior)	-	.00	.58
EX	11.53	.61	1.70
W (m = 4)	.86	-	-

(*) most favourable choice of i

No importance function is uniformly more efficient; conversely PTDC is always the least efficient.

2) On the basis of our experiments, we recommend the following "strategy" in the selection of an importance function :

i) one should first try STUD with a limited number[15] of drawings and a look at intermediate results to check if the estimated variation coefficient of the reciprocal of the integrating constant of the posterior density does not increase with the number of drawings.

ii) if this is the case, one should consider PTFC if the correlations (estimated by using STUD) between the coefficients of the different equations are close to 0., or PTST-i if that is not the case. The choice of i has been discussed in section III.2.5.

In all cases it is worthwhile to try a second round to see if a significant increase of relative precision is obtained.

3) One should not be too optimistic about the size of models that can be treated. Here the relevant measure of size is given by the number of structural coefficients (ℓ); the number of stochastic equations is also relevant for PTFC and PTDC. However, even in the classical approach, there are not so many models which are estimated by FIML, simply because there are not enough observations. In such a case, one has to use limited information methods, or to assume block recursivity in the system. The Bayesian approach can be applied then to each subsystem, which fits in the framework we have described in this study.

CHAPTER V : EXTENSIONS

V.1 *PRIOR DENSITY*

We have considered the case of the (truncated) extended natural-conjugate prior. Its drawbacks are well known - see DR and Richard (1973). In particular, one must elicit an informative prior on Σ, a difficult task; in addition, the prior information on Σ has a strong influence on the results on δ, and especially on β. However, we can extend our approach to the case of the independent Student-inverted-Wishart prior :

(5.1) $$p(\Sigma, \delta) = p(\Sigma) \quad p(\delta)$$

where $p(\Sigma) = f^{m}_{iW} \; (\Sigma \mid S_0, \nu_0)$, or

$$\propto |\Sigma|^{-\frac{1}{2}(\nu_0+m+1)}$$

$$p(\delta) = f^{\ell}_{t}(\delta \mid \delta_0, P_0, \lambda_0);$$

in particular $p(\delta)$ can take the form (2.4), including DR's invariant prior with respect to the normalization rule.

With the prior (5.1), one has to replace 1-1 poly-t densities by 2-1 poly-t densities in all the poly-t based importance functions of section III.2. The additional Student kernel comes from the conditional prior $p(\delta_i \mid \delta_{\bar{i}})$.

It is technically possible to generate random drawings from 2-1 poly-t densities, see Appendix A, but it is more costly than to generate random drawings from 1-1 poly-t densities[16]. Furthermore, there is no reason why the poly-t based importance functions would perform worse with the prior (5.1) than with the extended natural-conjugate prior. This suggests that it may be worthwhile to implement importance functions based on 2-1 poly-t densities when the prior is (5.1).

V.2 *NONLINEAR MODELS*

We distinguish 2 cases :

1) The stochastic equations are linear in the parameters but some identities are nonlinear; so that the model (1.1) becomes (at period t) :

(5.2) $$\tilde{y}'_t B_E + z'_t \Gamma_E = (y'_t \quad y'_{It}) \begin{pmatrix} B_{EE} \\ B_{EI} \end{pmatrix} + z'_t \Gamma_E = u'_t$$

(5.3) $$h'(\tilde{y}'_t, z'_t) = h'(y'_t, y'_{It}, z'_t) \qquad = 0'$$

where B_E is $\tilde{m} \times m$, B_{EE} is $m \times m$, Γ_E is $n \times m$ and h is $1 \times (\tilde{m} - m)$, i.e. a vector of functions corresponding to the identities. The partitioning of $\tilde{y}_t'$ into y_t' and y_{It}' may not be unique. In the likelihood function (1.7), the Jacobian factor $\|B\|^T$ is replaced by

$$(5.4) \qquad \prod_{t=1}^{T} \|J_t\| = \prod_{t=1}^{T} \|B_{EE} - B_{EI} B_{IIt}^{-1} B_{IEt}\|$$

where

$$(5.5) \qquad B_{IIt} = \left(\frac{\partial h'}{\partial y_{It}}\right), \text{ of order } \tilde{m} - m \text{ and } B_{IEt} = \left(\frac{\partial h'}{\partial y_t}\right) \text{ of dimensions } (\tilde{m} - m) \times m$$

See Dagenais (1978, section 1.5) for the general case where (5.2) is also a set of nonlinear functions.

In the linear case, the Jacobian $\|B\|^T$ of (1.7) is the cause of the presence of the quadratic form Q_{0i} in the conditional posterior density $p(\delta_i \mid \delta_{\bar{i}})$, see (3.3). In the nonlinear case, the factor arising from (5.4) is not quadratic and in general, it is therefore no longer possible to express this density as a 1-1 poly-t. However, let B_t denote the matrix

$$(5.6) \qquad B_t = \begin{bmatrix} B_{EE} & B_{IEt}' \\ B_{EI} & B_{IIt} \end{bmatrix}$$

As $|J_t| = |B_t| \cdot |B_{IIt}|^{-1}$, we can express (5.4) as

$$(5.7) \qquad \prod_{t=1}^{T} |B_t| \cdot |B_{IIt}|^{-1}$$

and neglect the factor $\prod_{t=1}^{T} |B_{IIt}|^{-1}$ to form the conditional posterior density, since this factor does not depend on δ.

By applying formula (B.1) to $B_t = (\bar{\beta}_1 \quad B_{\bar{1}t})$, the conditional posterior density $p(\delta_1 \mid \delta_{\bar{1}})$ is given by

$$(5.8) \qquad p(\delta_1 \mid \delta_{\bar{1}}) \propto \prod_{t=1}^{T} (\bar{\beta}_1' N_t \bar{\beta}_1)\, Q_{11}^{-\lambda_1} = \prod_{t=1}^{T} Q_{01t}\, Q_{11}^{-\lambda_1}$$

where $N_t = I_m - B_{\bar{1}t} (B_{\bar{1}t}' B_{\bar{1}t})^{-1} B_{\bar{1}t}'$ and Q_{01t} is a quadratic form in the parameters β_1 of $\bar{\beta}_1$ obtained by reexpressing $\bar{\beta}_1' N_t \bar{\beta}_1$ adequately. The density (5.8) is a $T-1$ poly-t density in $\delta_1 = (\beta_1' \; \gamma_1')'$. Since Q_{01t} is not a function of γ_1 (the parameters of the predetermined variables of equation 1), (5.8) can be factorized into

(5.9) $\quad p(\beta_1 \mid \delta_{\bar{1}})\, p(\gamma_1 \mid \beta_1, \delta_{\bar{1}})$

where $p(\beta_1 \mid \delta_{\bar{1}})$ is still a $T-1$ poly-t and $p(\gamma_1 \mid \beta_1, \delta_{\bar{1}})$ is a Student density. $p(\beta_1 \mid \delta_{\bar{1}})$ could be used as such in the poly-t based importance functions of section III.2; indeed it could then still be possible to generate random drawings from it at "reasonable" cost, e.g. by a rejection method based on drawings from a 1-1 poly-t with kernel $Q_{01t}\, Q_{11}^{-\lambda_1}$, or by tabulation of the distribution function if the dimension of β_1 is small (1 or 2). If this cannot be done at reasonable cost, $p(\beta_1 \mid \delta_{\bar{1}})$ has to be replaced by an importance function, e.g. a 1-1 poly-t with kernel $\overline{Q}_{01}^{\frac{1}{2}T}\, Q_{11}^{-\lambda_1}$ where the parameters of $\overline{Q}_{01}$ are suitable functions of those of Q_{01t} $(t = 1,\ldots,T)$

One can also more simply linearize the identities (5.3) and replace (5.4) by $\|B\|^T$, as obtained from the linearized model, in order to use the corresponding (1-1) poly-t based importance functions of section III.2.

2) Some stochastic equations are nonlinear in the parameters : we could also build poly-t based importance functions using a linearized version of the model. For applications in other contexts, see Florens (1977) and Lubrano (1983).

CONCLUSION

In this monograph, we have tried to exploit as far as possible the analytical results derived by Drèze, Morales and Richard, in order to build the importance functions which can be used to carry out the integration required by a Bayesian analysis of the simultaneous equation model, or particular versions thereof, as was initially proposed by Kloek and van Dijk. We have proposed several automatic importance functions based on poly-t densities, and, in some cases, have thereby been able to achieve significant variance reduction. As is well known within the framework of Monte Carlo methods, variance reduction results from efficient use of any kind of relevant information, whether it be theoretical or empirical.

In this approach, we have restricted the prior density of δ and Σ to be in the extended natural-conjugate family or in the independent Student-inverted-Wishart family. That they need not (but can) be truncated with respect to δ may be an advantage since by truncation one might lose relevant sample information. Such information could be helpful for the search of a better specification of a model, since the way in which the prior information is modified by the sample can lead to change the specification. Nevertheless, it may happen that these classes of prior densities are too restrictive, although there are very few families of *multivariate* distributions that are flexible enough to represent prior information. Hence we suggest to use the two classes we have described above : as usual in the Bayesian approach to inference, there may be a bargain between the adequate representation of the prior information and the numerical efficiency.

Finally, we want to emphasize that we do not consider the approach developed in this research to be necessarily the best. More experiments are certainly needed to validate and improve it. An alternative approach is to use the importance functions of Kloek and van Dijk. It is conceivable that some of their methods could be combined with ours, to obtain still better importance function (in particular by refining the importance functions $f(\delta_{\bar{i}})$ of PTST-i, see (3.17)).

APPENDIX A : DENSITY FUNCTIONS : DEFINITIONS, PROPERTIES AND ALGORITHMS FOR GENERATING RANDOM DRAWINGS

We give here the definitions of the probability density functions used in this book, i.e.

- the matricvariate normal, and its particular case, the multivariate normal, (section A.I),
- the inverted-Wishart, and its particular case, the inverted-gamma (section A.II),
- the multivariate Student (section A.III),
- the 2-0 poly-t (section A.IV),
- the m-1 poly-t, for $m = 1$ or 2 (section A.V).

For each class, we also state a few properties of the density and give an algorithm to generate random drawings from it.

Each section of this appendix can be read independently of the others. The following notation is used throughout several sections :

- C_p $(\overline{C}_p)$ denotes the set of $p \times p$ PDS (PSDS) matrices,
- LT means lower triangular,
- $\Gamma(\alpha) = \int_0^\infty x^{\alpha-1} \exp - x \, dx$ $(\alpha > 0)$ is the gamma function,
- $Q_{\bullet} = [s_{\bullet} + (y - \overline{y}_{\bullet})' P_{\bullet} (y - \overline{y}_{\bullet})]$ is a positive quadratic form in $y \in \mathbb{R}^p$, with parameters $s_{\bullet} \geqslant 0$, $\overline{y}_{\bullet} \in \mathbb{R}^p$ and $P_{\bullet} \in \overline{C}_p$.

A.I. *THE MATRICVARIATE NORMAL (MN) DISTRIBUTION*

1. *Definition* : the $p \times m$ random matrix $Y \in \mathbb{R}^{pm}$ has an MN distribution if and only if its density function is

$$\text{(A.1)} \quad p(Y) = f_{MN}^{p\times m}(Y \mid \overline{Y}, \Sigma \otimes P)$$

$$= [(2\pi)^{pm} |\Sigma|^p |P|^m]^{-\frac{1}{2}} \exp - \frac{1}{2} \operatorname{tr} \Sigma^{-1} (Y - \overline{Y}) P^{-1} (Y - \overline{Y})$$

where $\overline{Y} \in \mathbb{R}^{pm}$, $\Sigma \in C_m$ and $P \in C_p$.

Special case : the *multivariate* normal distribution is a special case of (A.1), namely $m = 1$, and is denoted :

(A.2) $$p(y) = f_N^p(y \mid \bar{y}, P) = [(2\pi)^p |P|]^{-\frac{1}{2}} \exp - \frac{1}{2}(y-\bar{y})' P^{-1}(y-\bar{y})$$

where $\bar{y} \in \mathbb{R}^p$ and $P \in C_p$.

2. *Properties*

(i) $E(Y) = \bar{Y}$ and $V(Y) = \Sigma \otimes P$; $V(Y)$ denotes the variance-covariance matrix of the column expansion of Y.

(ii) If A is an $r \times p$ matrix of rank $r \leqslant p$, and B is an $m \times t$ matrix of rank $t \leqslant m$, then

(A.3) $$p(AYB) = f_{NM}^{r\times t}(AYB \mid A\bar{Y}B,\ B'\Sigma B \otimes APA').$$

For additional properties and proofs, see Richard (1979).

3. The following algorithm can be used to generate a matrix Y from (A.1) :

MN ALGORITHM

1. Compute the LT matrices B' and A such that $\Sigma = B'B$ and $P = AA'$.
2. Generate mp independent univariate standard normal drawings, e.g. by using the polar algorithm - see Knuth (1971) - and put them in a $p \times m$ matrix Z.
3. Compute $Y = AZB + \bar{Y}$.

Step 2 amounts to generate a drawing from $p(Z) = f_{MN}^{p\times m}(Z \mid 0, I_m \otimes I_p)$.
Step 3 follows from an application of (A.3) to $p(Z)$. Note that it is more efficient to choose B' and A from the triangular decomposition of Σ and P than from their spectral decomposition.

A.II *THE INVERTED-WISHART (iW) DISTRIBUTION*

1. *Definition* : the $m \times m$ random matrix $\Sigma \in C_m$ has an iW distribution if and only if its density function is

(A.4) $$p(\Sigma) = f_{iW}^m(\Sigma \mid S, \nu) = \left[2^{\frac{1}{2}\nu m}\, \pi^{\frac{1}{4}m(m-1)} \prod_{i=1}^{m} \Gamma\left(\frac{\nu+1-i}{2}\right)\right]^{-1} |S|^{\frac{1}{2}\nu}\, |\Sigma|^{-\frac{1}{2}(\nu+m+1)} \exp -\frac{1}{2}\operatorname{tr}\Sigma^{-1}S$$

where $S \in C_m$ and $\nu > m-1$.

Special case : the inverted-gamma distribution corresponds to the case where $m = 1$

and is denoted

(A.5) $$p(\sigma^2) = f_{i\gamma}(\sigma^2 \mid s, \nu)$$

$$= [\Gamma(\tfrac{\nu}{2})]^{-1} (\tfrac{s}{2})^{\frac{1}{2}\nu} (\sigma^2)^{-\frac{1}{2}(\nu+2)} \exp - \frac{1}{2}\frac{s}{\sigma^2}$$

where $s > 0$ and $\nu > 0$.

2. *Properties*

(i) If C is an $m \times r$ matrix of rank r and $\Omega = C'\Sigma C$, then

(A.6) $$p(\Omega) = f^{r}_{iW}(\Omega \mid C'SC, \nu - m + r).$$

(ii) If Σ is partitioned into $\Sigma_{11} \in C_{m_1}$, $\Sigma_{22} \in C_{m_2}$, Σ_{12} and Σ_{21}, and $\Sigma_{22.1} = \Sigma_{22} - \Sigma_{21}\Sigma_{11}^{-1}\Sigma_{12}$, then

(A.7) $$p(\Sigma_{11}, \Sigma_{11}^{-1}\Sigma_{12}, \Sigma_{22.1}) = p(\Sigma_{11})\, p(\Sigma_{11}^{-1}\Sigma_{12} \mid \Sigma_{22.1})\, p(\Sigma_{22.1})$$

with

(A.8) $$p(\Sigma_{11}) = f^{m_1}_{iW}(\Sigma_{11} \mid S_{11}, \nu - m_2)$$

(A.9) $$p(\Sigma_{11}^{-1}\Sigma_{12} \mid \Sigma_{22.1}) = f^{m_1 \times m_2}_{MN}(\Sigma_{11}^{-1}\Sigma_{12} \mid S_{11}^{-1}S_{12}, \Sigma_{22.1} \otimes S_{11}^{-1})$$

(A.10) $$p(\Sigma_{22.1}) = f^{m_2}_{iW}(\Sigma_{22.1} \mid S_{22.1}, \nu)$$

where S_{11}, S_{12} and $S_{22.1}$ are defined from S as Σ_{11}, Σ_{12} and $\Sigma_{22.1}$ are defined from Σ. For the meaning of (A.9), see (A.1).

(iii) If $\nu > m + 1$,

$$E(\Sigma) = \frac{1}{\nu - m - 1} S.$$

For proofs, see Richard (1979).

3. The following algorithm can be used to generate a matrix Σ from (B.4) :

iW ALGORITHM

1. Compute the LT matrix C such that $S = CC'$.
   $k \leftarrow 0$; $\ell \leftarrow \frac{1}{2} m(m+1) + 1$; dimension a vector w of ℓ elements.

2. $k \leftarrow k+1$; if $k > m$, go to 7.

3. Generate a drawing from $f_{i\gamma}(w_k \mid 1, \nu - k + 1)$;

   $\ell \leftarrow \ell - k$;

   $w(\ell) \leftarrow \sqrt{w_k}$.

4. If $k = 1$, go to 2; else, continue.

5. Generate a vector z of $k-1$ independent univariate standard normal drawings.

   Compute $y = \sqrt{w_k} \cdot W_{k-1} \cdot z$, where W_{k-1} denotes the LT matrix whose column expansion (neglecting the 0's above the main diagonal) is stored in the last $\frac{1}{2}k(k-1)$ elements of the vector w.

   Assign : $w(\ell + i) \leftarrow y(i)$, $i = 1,2,\ldots,k-1$.

6. Go to 2.

7. Compute $\Sigma = C W W' C'$, where W denotes the LT matrix stored in w (in the same way as W_{k-1} at step 5).

One must generate m inverted-gamma drawings (see step 3) - using e.g. the GRUB algorithm of Kinderman and Monahan (1980) - and $1 + 2 + \ldots + (m-1) = \frac{1}{2} m(m-1)$ standard normal drawings (see step 5).

Step 1 reduces the problem of drawing from (A.4) to that of drawing from $p(\Omega) = f_{iW}^{m}(\Omega \mid I_m, \nu)$, as follows from (A.6) and $S = CC'$.

Step 7 performs the inverse transformation $\Sigma = C \Omega C'$, for $\Omega = WW'$ where W is LT. Note that $V = CW$ is the LT decomposition of Σ.

Steps 2 to 6 perform the drawing of W in a recursive way, using (A.8) - (A.10). Indeed, notice that when $S = I_m$, $S_{22.1} = I_{m_2}$ so that drawing from (A.10) is the same problem as drawing from (A.4), except that the dimension has been reduced.

The advantage of drawing the LT matrix W such that $WW' = \Omega$ comes from the verification that $\Omega_{22.1} = W_{22} W_{22}'$, where W_{22} is selected from W as Σ_{22} from Σ : take e.g. $m_1 = 1$:

$$\text{(A.11)} \quad \begin{bmatrix} \omega_{11} & \Omega_{12} \\ \Omega_{11} & \Omega_{22} \end{bmatrix} = \begin{bmatrix} w_{11} & 0 \\ W_{21} & W_{22} \end{bmatrix} \begin{bmatrix} w_{11} & W_{21}' \\ 0 & W_{22}' \end{bmatrix} = \begin{bmatrix} w_{11}^2 & w_{11} W_{21}' \\ W_{21} w_{11} & W_{21} W_{21}' + W_{22} W_{22}' \end{bmatrix}$$

wherefrom

$$\text{(A.12)} \quad \omega_{11}^{-1} \Omega_{12} = w_{11}^{-1} W_{12}$$

$$\text{(A.13)} \quad \Omega_{22.1} = W_{22} W_{22}'$$

It follows then from (A.11) and (A.9) that

(A.14) $$p(w_{11}^{-1}W_{12} \mid W_{22}) = f_{MN}^{1\times m_2}(w_{11}^{-1}W_{12} \mid 0, W_{22}W_{22}').$$

Hence, W_{22} is the LT decomposition of the covariance matrix of this conditional density and is *directly* available for computing y at step 5 (where it is denoted W_{k-1}).

This algorithm can be easily adapted to the case of the Wishart distribution, for which a decomposition like (A.7) exists. In this case, it differs slightly from the algorithm presented by Kennedy and Gentle (1980, p. 231) and due to Odell and Feiveson (1966).

A.III. *THE MULTIVARIATE STUDENT DISTRIBUTION*

1. *Definition* : the $p\times 1$ random vector y has a multivariate Student distribution if and only if its density function is

(A.15) $$p(y) = f_t^p(y \mid \bar{y}, \frac{P}{s}, \nu)$$

$$[\pi^{\frac{1}{2}p}\,\Gamma(\frac{\nu}{2})/\Gamma(\frac{\nu+p}{2})]^{-1}\; s^{\frac{1}{2}\nu}\,|P|^{\frac{1}{2}}\,Q^{-\frac{1}{2}(\nu+p)}$$

where $s > 0$, $\bar{y} \in \mathbb{R}^p$ and $P \in C_p$ are the parameters of the quadratic form Q.

2. *Properties*

(i) $E(y) = \bar{y} \quad (\nu > 1)$

$$V(y) = \frac{1}{\nu-2}P^{-1} \quad (\nu > 2).$$

(ii) Moments of y exist up to the order ν (i.e. ν excluded).

(iii) If $p(y \mid \sigma^2) = f_N^p(y \mid \bar{y}, \sigma^2 P^{-1})$ - see (A.2) and $p(\sigma^2) = f_{i\gamma}(\sigma^2 \mid s, \nu)$ - see (A.5) -, then

(A.16) $$p(y) = f_t^p(y \mid \bar{y}, \frac{P}{s}, \nu)$$

(A.17) $$p(\sigma^2 \mid y) = f_{i\gamma}(\sigma^2 \mid Q, \nu+p).$$

For other properties and proofs, see Richard (1979).

3. The following algorithm can be used to generate a Student vector y from (A.15) :

STUDENT ALGORITHM

1. Generate an inverted-gamma drawing from $f_{i\gamma}(\sigma^2 \mid s, \nu)$.
2. Generate a normal drawing from $f^p(y \mid \bar{y}, \sigma^2 P^{-1})$, where σ^2 is the value drawn at step 1.

It is evidently an application of property (iii) above.

A.IV. *The 2-0 POLY-T DISTRIBUTION*

1. *Definition* : the $p \times 1$ random vector y has a 2-0 poly-t distribution if and only if its density function has a kernel given by

(A.18) $$\varphi_2^p(y \mid S) = Q_1^{-\frac{1}{2}\nu_1} Q_2^{-\frac{1}{2}\nu_2}$$

with parameters $S = (\{s_j, \bar{y}_j, P_j, \nu_j\}\ ;\ j = 1,2)$ subject to the constraints

(A.19a) $$s_j > 0, \quad \bar{y}_j \in \mathbb{R}^p$$

(A.19b) $$P_j \in \bar{C}_p, \quad \sum_{j=1}^{2} P_j \in C_p$$

(A.19c) $$\nu_j > 0, \quad \nu = \sum_{j=1}^{2} \nu_j - p > 0.$$

2. *Properties* : the following identity holds :

(A.20) $$\varphi_2^p(y \mid S) = \int_0^1 g(c \mid S)\, f_t^p(y \mid \bar{y}_c, s_c^{-1} \cdot P_c, \nu)\, dc$$

where

(A.21) $$g(c \mid S) = \pi^{\frac{1}{2}p} \frac{\Gamma\left(\frac{\nu_1 + \nu_2}{2}\right)}{\Gamma(\frac{1}{2}\nu_1)\,\Gamma(\frac{1}{2}\nu_2)} [|P_c| \cdot s_c]^{-\frac{1}{2}}\, c^{\frac{1}{2}\nu_1 - 1}\, (1-c)^{\frac{1}{2}\nu_2 - 1}$$

(A.22a) $$P_c = c P_1 + (1-c) P_2$$

(A.22b) $$s_c = c(s_1 + \bar{y}_1' P_1 \bar{y}_1) + (1-c)(s_2 + \bar{y}_2' P_2 \bar{y}_2) - \bar{y}_c' P_c \bar{y}_c$$

(A.22c) $$\bar{y}_c = P_c^{-1} [c P_1 \bar{y}_1 + (1-c) P_2 \bar{y}_2]$$

For a proof of (A.20), see Richard and Tompa (1980).

Therefore, the reciprocal of the integrating constant of a 2-0 poly-t can be expressed as

$$\text{(A.23)} \qquad K = \int_{\mathbb{R}^p} \varphi_2^p (y \mid S)\, dy = \int_0^1 g(c \mid S)\, dc$$

and the expected value of any integrable function $g(y)$ as

$$\text{(A.24)} \qquad E[g(y)] = \frac{1}{K} \int_0^1 E[g(y) \mid c] \cdot g(c \mid S)\, dc$$

where $E[g(y) \mid c]$ is the conditional expectation of $g(y)$, y having the Student density $f_t^p (y \mid \bar{y}_c, s_c^{-1} \cdot P_c, \nu)$; e.g. for $g(y) = y$, it is $\bar{y}_c$.

By property (ii) of the Student distribution, moments of y exist up to the order ν.

The integrals of the right-hand sides of (A.23) and (A.24) must be computed by numerical integrations with respect to c.

3. The following algorithm can be used to generate a random vector y from a 2-0 poly-t density :

<u>2-0 POLY-T ALGORITHM</u>

1. Generate a drawing from the density $p(c) = K^{-1} \cdot g(c \mid S)$ - see (A.21) and (A.23) -, by a standard method, such as a rejection method or the direct method of inverting the distribution function.
2. Generate an inverted-gamma drawing from $f_{i\gamma} (\sigma^2 \mid s_c, \nu)$, where s_c is computed using the value of c obtained at step 1 - see (A.22b).
3. Generate a normal drawing from $f_N^p (y \mid \bar{y}_c, \sigma^2 P_c^{-1})$, where $\bar{y}_c$ and P_c are computing using the value of c obtained at step 1 - see (A.22a) and (A.22c) - and σ^2 is the value drawn at step 2.

This algorithm is evidently based on (A.20) and the Student algorithm. It can be used as such when one needs only a *very limited* number, say 10 at most, of drawings. This is the case when one uses the poly-t based importance functions of section III.2 for computing posterior moments of the coefficients of a "seemingly unrelated" regression model, analysed with an independent Student-inverted-Wishart prior density; see section V.1 and footnote 16. Indeed, in that case, the 2-0 poly-t densities used in the importance functions are densities for the coefficients of one equation, conditional on the coefficients of the other equations; their parameters are functions of these coefficients and therefore change at each drawing.

For a variant of this algorithm better suited to the case where many random drawings of the same 2-0 poly-t are needed, see Bauwens and Richard (1980).

A.V. THE m-1 ($0 < m \leq 2$) POLY-T DISTRIBUTION

1. *Definition* : the $p \times 1$ random vector y has an $m-1$ poly-t distribution if and only if its density function has a kernel given by

$$\text{(A.25)} \qquad \varphi^{p}_{m,1}(y \mid S_0, S) = Q_0^{\frac{1}{2}\nu_0} \prod_{j=1}^{m} Q_j^{-\frac{1}{2}\nu_j}$$

with parameters $S = (\{s_j, \bar{y}_j, P_j, \nu_j\};\ j = 1,\dots,m)$ subject to the constraints (A.19a) - (A.19c) (with m substituted for 2) and $S_0 = \{s_0, \bar{y}_0, P_0, \nu_0\}$ subject to the constraints

$$s_0 \geq 0, \quad \bar{y}_0 \in \mathbb{R}^p, \quad P_0 \in \bar{\mathcal{C}}_p, \quad \nu_0 > 0$$

and

$$\sum_{j=1}^{m} \nu_j - \nu_0 - p = \nu - \nu_0 > 0. \quad \text{Let } \ell_0 = \frac{1}{2}\nu_0.$$

Special case : for $m = 1$, we define the canonical form of the 1-1 poly-t as the case where $s_1 = 1$, $\bar{y}_1 = 0$, $P_1 = I_p$, $\bar{y}_0 = z_0$ and $P_0 = \Lambda_0 = \text{diag}(\lambda_{01}\dots\lambda_{0p})$ and denote it

$$\text{(A.26)} \qquad \tilde{\varphi}_{1,1}(z) = [s_0 + (z-z_0)'\Lambda_0(z-z_0)]^{\frac{1}{2}\nu_0}\,[1+z'z]^{-\frac{1}{2}\nu_1}$$

$$= \bar{Q}_0^{\ell_0}\,\bar{Q}_1^{-\frac{1}{2}\nu_1}.$$

2. *Properties*

(i) We shall give the properties of the 1-1 poly-t (i.e. $m = 1$) since by using (A.20) the 2-1 poly-t can be expressed as - see Richard and Tompa (1980) -

$$\text{(A.27)} \qquad \varphi^{p}_{2,1}(y \mid S_0, S) = Q_0^{\ell_0}\int_0^1 Q_c^{-\frac{1}{2}(\nu_1+\nu_2)}\, f_\beta(c \mid \tfrac{1}{2}\nu_1, \tfrac{1}{2}\nu_2)\,dc$$

$$= \int_0^1 \varphi^{p}_{1,1}(y \mid S_0, S_c)\cdot f_\beta(c \mid \tfrac{1}{2}\nu_1, \tfrac{1}{2}\nu_2)\,dc$$

where

$$f_\beta(c \mid \nu_1, \nu_2) = \frac{\Gamma\left(\frac{\nu_1+\nu_2}{2}\right)}{\Gamma(\frac{1}{2}\nu_1)\,\Gamma(\frac{1}{2}\nu_2)}\, c^{\frac{1}{2}\nu_1-1}\,(1-c)^{\frac{1}{2}\nu_2-1}$$

and

$$S_c = \{s_c, \bar{y}_c, P_c, \nu_1+\nu_2\}, \text{ see (A.22).}$$

Therefrom, the reciprocal of the integrating constant of a 2-1 poly-t is given by

$$\text{(A.28)} \qquad K_{2,1} = \int_{\mathbb{R}^p} \varphi_{2,1}(y \mid S_0, S)\,dy = \int_0^1 K_{1,1}(c)\cdot f_\beta(c \mid \tfrac{1}{2}\nu_1, \tfrac{1}{2}\nu_2)\,dc$$

where

$$\text{(A.29)} \qquad K_{1,1}(c) = \int_{\mathbb{R}^p} \varphi^p_{1,1}(y \mid S_0, S_c)\, dy$$

Similarly, for any integrable function $g(y)$,

$$\text{(A.30)} \qquad E[g(y)] = K^{-1}_{2,1} \int_0^1 E[g(y) \mid c] \cdot K_{1,1}(c) \cdot f_\beta\left(c \mid \tfrac{1}{2}\nu_1, \tfrac{1}{2}\nu_2\right) dc$$

where $E[g(y) \mid c]$ is the conditional expectation of $g(y)$, y having the 1-1 poly-t distribution with kernel $\varphi^p_{1,1}(y \mid S_0, S_c)$.

The integrals of the right-hand sides of (A.28) and (A.30) must be computed numerically.

(ii) Moments of y exist up to the order r if $\sum_{j=1}^{m} \nu_j - \nu_0 - p > r$.

(iii) Moments of the 1-1 poly-t density *for integer* ℓ_0 can be computed analytically by recursion formulas given by Richard and Tompa (1980) or by formulas given by Bauwens and Richard (1982, Appendix).

(iv) A 1-1 poly-t density can be reduced to a canonical form (A.26) by a linear transformation

$$\text{(A.31)} \qquad z = T^{-1}(y - \bar{y}_1)$$

where T is the nonsingular matrix that diagonalizes simultaneously P_1 and P_0, i.e. $T'P_1T = s_1 I_p$ and $T'P_0T = \Lambda_0 = \operatorname{diag}(\lambda_{01} \dots \lambda_{0p})$; moreover :

$$\text{(A.32)} \qquad z_0 = T^{-1}(\bar{y}_0 - \bar{y}_1)\,.$$

Note that, more generally, a linear transformation of a vector y having an m-1 poly-t distribution is a vector in the same class of distributions.

(v) The following identity holds, by property (iii) of the Student distribution :

$$\text{(A.33)} \qquad \tilde{\varphi}_{1,1}(z) \propto \int_{\mathbb{R}_+} \bar{Q}_0^{\ell_0} \cdot f^p_N\left(z \mid 0, \tfrac{1}{2}\sigma^2 I_p\right) \cdot f_{i\gamma}(\sigma^2 \mid 2, \nu_1 - p)\, d\sigma^2$$

wherefrom

$$\text{(A.34)} \qquad p(z \mid \sigma^2) \propto \bar{Q}_0^{\ell_0} \exp - \frac{z'z}{\sigma^2}$$

Note that (A.33) could be formulated more generally for the case of $\varphi_{1,1}(y \mid S_0, S)$.

The density $p(z \mid \sigma^2)$ is a quadratic form in normal variables. Assuming ℓ_0 to be integer, it can be expressed as a finite mixture of products of independent univariate densities whose integrating constants are reciprocals of moments in normal variables :

$$\text{(A.35)} \qquad p(z \mid \sigma^2) = \sum_{n \in N} P[N = n \mid \sigma^2] \prod_{j=1}^{p} p(z_j \mid n_j, z_{0j}, \sigma^2)$$

where

N is a $p+1$ dimensional discrete random vector with frequency function $P[N = n \mid \sigma^2]$ defined below

$$N = \{n : 0 \leqslant n_j \leqslant \ell_0, \; j = 1 \to p; \; n_{p+1} = \ell_0 - \sum_{j=1}^{p} n_j; \; n_{p+1} \equiv 0 \text{ if } s_0 = 0\}$$

$$\text{(A.36)} \qquad p(z_j \mid n_j, z_{0j}, \sigma^2) = S_j^{-1} \cdot (z_j - z_{0j})^{2n_j} \exp - \frac{z_j^2}{\sigma^2}$$

$$\text{(A.37)} \qquad S_j = S_j(n_j, z_{0j}, \sigma^2) = \int_{\mathbb{R}} (z_j - z_{0j})^{2n_j} \exp - \frac{z_j^2}{\sigma^2} \, dz_j$$

$$\text{(A.38)} \qquad P[N = n \mid \sigma^2] = C^{-1}(\sigma^2) \cdot \alpha_n \cdot s_0^{n_{p+1}} \cdot \prod_{j=1}^{p} \lambda_{0j}^{n_j} S_j$$

$$\text{(A.39)} \qquad C(\sigma^2) = \sum_{n \in N} \alpha_n \cdot s_0^{n_{p+1}} \cdot \prod_{j=1}^{p} \lambda_{0j}^{n_j} S_j$$

$$\alpha_n = \frac{\ell_0!}{n_1! \ldots n_{p+1}!}$$

As $C^{-1}(\sigma^2)$ is the integrating constant of $p(z \mid \sigma^2)$, the marginal density of σ^2 has a kernel given by

$$\text{(A.40)} \qquad p(\sigma^2) \propto C(\sigma^2) \cdot f_{i\gamma}(\sigma^2 \mid 2, \nu_1 - p).$$

It is shown in Bauwens and Richard (1982) that the integrating constant of $p(\sigma^2)$ can be obtained analytically, as indeed the integrals (A.37) are known, since they are moments in normal variables.

3. The following algorithm can be used for generating random drawings from a 1-1 poly-t density with kernel $\varphi^{p}_{1,1}(y \mid S_0, S)$ and integer ℓ_0 :

<u>1-1 POLY-T ALGORITHM (integer ℓ_0)</u>

1. Reduce $\varphi^{p}_{1,1}(y \mid S_0, S)$ to its canonical form (A.26).
2. Generate a random drawing σ^2 from $p(\sigma^2)$ as given by (A.40).

3. Generate a random drawing from $p(z \mid \sigma^2)$ where σ^2 is the value obtained at step 2, i.e.

   a) generate the discrete vector N from its distribution defined by $P[N = n \mid \sigma^2]$, see (A.38);

   b) generate z_j from $p(z_j \mid n_j, z_{0j}, \sigma^2)$ - see (A.36) - where n_j is the j-th element of the vector N generated at step 3.a).

4. Transform z into $y = Tz + \bar{y}_1$.

Steps 1 and 4 follow immediately from property (iv), while *steps 2 and 3* operate a sequential random drawing procedure. Their implementation is discussed in detail by Bauwens and Richard (1982).

If ℓ_0 is not integer, a rejection procedure based on drawings from a 1-1 poly-t with ℓ_0 replaced by its ceil L_0 is proposed by these authors.

Finally, by (A.27), a random drawing from a 2-1 poly-t with integer ℓ_0 can be obtained by the following algorithm :

<u>2-1 POLY-T ALGORITHM (integer ℓ_0)</u>

1. Generate a random drawing c from p(c) whose kernel is given by $K_{1,1}(c) \cdot f_\beta(c \mid \frac{1}{2}\nu_1, \frac{1}{2}\nu_2)$ - see (A.29).

2. Apply the 1-1 poly-t algorithm to draw y from the 1-1 poly-t with kernel $\varphi_{1,1}(y \mid S_0, S_c)$ - see (A.27).

APPENDIX B : THE TECHNICALITIES OF CHAPTER III

B.I. *DEFINITION OF THE PARAMETERS OF (3.3) AND (3.6)*

For convenience of notation, we give the formulas of the parameters of (3.3) and (3.6) for the case $i = 1$.

The parameters of Q_{01} are functions of a subset of the conditioning values $\delta_{\bar{1}}$ only. This quadratic form is obtained from the factor $\|B\|$ of (3.1). Let $B = [\bar{\beta}_1 \quad B_{\bar{1}}]$ where $\bar{\beta}_1$ is the first column of B. Then

$$\text{(B.1)} \qquad \|B\| = |B'B|^{\frac{1}{2}} = [(\bar{\beta}_1' N \bar{\beta}_1) | B_{\bar{1}}' B_{\bar{1}}|]^{\frac{1}{2}}$$

where $\qquad N = I_m - B_{\bar{1}} (B_{\bar{1}}' B_{\bar{1}})^{-1} B_{\bar{1}}'$.

The first factor of the right-hand side of (B.1) is a quadratic form in $\bar{\beta}_1$; since $\bar{\beta}_1$ contains free parameters, say β_1, 0's and one element equal to 1 (by the normalisation rule), this quadratic form in $\bar{\beta}_1$ can be expressed as a quadratic form in the free parameters β_1, and therefore also in δ_1, since β_1 is included in δ_1. The conditioning values affect this quadratic form through the free parameters of $B_{\bar{1}}$. Note that the matrix N is of rank 1.

The parameters of Q_{11} are functions of $\delta_{\bar{1}}$, M_*, Δ_* and S_*. Q_{11} comes from the factor $|S_* + (\Delta - \Delta_*)' M_* (\Delta - \Delta_*)|$ of (3.1). Indeed, this determinant can be factorized into a product of a quadratic form in δ_1 and a determinant of a matrix that does not depend on δ_1, in the same spirit as in (B.1) :

$$\text{(B.2)} \qquad |S_* + (\Delta - \Delta_*)' M_* (\Delta - \Delta_*)| = Q_{11} \, |\bar{S}_1 + (\Delta_{\bar{1}} - \bar{\Delta}_1)' \bar{M}_1 (\Delta_{\bar{1}} - \bar{\Delta}_1)|.$$

The proof is given in DR (Appendix B, § 5 and 6).

Let us first define the parameters $\bar{s}_{11}, \bar{y}_{11}$ and $\bar{A}_{11}$ of the quadratic form $Q_{11} = \bar{s}_{11} + (\delta_1 - \bar{y}_{11})' A_{11} (\delta_1 - \bar{y}_{11})$.

We need to partition the matrices Δ, Δ_*, M_* and S_* as follows :

$$\Delta = \begin{bmatrix} \delta_1 & 0 \\ 0 & \Delta_{\bar{1}} \end{bmatrix}, \qquad \Delta_* = \begin{bmatrix} \delta_{*11} & \Delta_{*12} \\ \Delta_{*21} & \Delta_{*22} \end{bmatrix}$$

$$S_* = \begin{bmatrix} s_{*11} & s_{*12} \\ s_{*21} & S_{*22} \end{bmatrix} \qquad M_* = \begin{bmatrix} M_{*11} & M_{*12} \\ M_{*21} & M_{*22} \end{bmatrix}$$

Then,

(B.3) $$\bar{s}_{11} = s_{*11.2} + [\Delta_{*21} + (\Delta_{\bar{1}} - \Delta_{*22}) S^{-1}_{*22}\, s_{*21}]' R_{22.1} [\Delta_{*21} + (\Delta_{\bar{1}} - \Delta_{*22}) S^{-1}_{*22}\, s_{*21}]$$

(B.4) $$\bar{y}_{11} = \delta_{*1} - \Delta_{*12} S^{-1}_{*22}\, s_{*21} + \bar{A}^{-1}_{11} R_{12} [\Delta_{*21} + (\Delta_{\bar{1}} - \Delta_{*22}) S^{-1}_{*22}\, s_{*21}]$$

(B.5) $$\bar{A}_{11} = M_{*11} - [M_{*11} \;\; M_{*12}] \begin{bmatrix} \Delta_{*12} \\ \Delta_{*22} - \Delta_{\bar{1}} \end{bmatrix} [\bar{S}_1 + (\Delta_{\bar{1}} - \bar{\Delta}_1)' \bar{M}_1 (\Delta_{\bar{1}} - \bar{\Delta}_1)]^{-1} \begin{bmatrix} \Delta_{*12} \\ \Delta_{*22} - \Delta_{\bar{1}} \end{bmatrix}' \begin{bmatrix} M_{*11} \\ M_{*21} \end{bmatrix}$$

where $$R = \begin{bmatrix} R_{11} & R_{12} \\ R_{21} & R_{22} \end{bmatrix} = M_* - M_* \begin{bmatrix} \Delta_{*12} \\ \Delta_{*22} - \Delta_{\bar{1}} \end{bmatrix} [\bar{S}_1 + (\Delta_{\bar{1}} - \bar{\Delta}_1)' \bar{M}_1 (\Delta_{\bar{1}} - \bar{\Delta}_1)]^{-1} \begin{bmatrix} \Delta_{*12} \\ \Delta_{*22} - \Delta_{\bar{1}} \end{bmatrix} M_*$$

and $$R_{22\cdot 1} = R_{22} - R_{21} R^{-1}_{11} R_{12}$$

Note that $\bar{A}_{11} = R_{11}$.

The parameters $\bar{S}_1$, $\bar{\Delta}_1$ and $\bar{M}_1$ of (B.2) and (B.5) are given by

(B.6) $$\bar{S}_1 = S_{*22} + \Delta'_{*12} M_{*11} \Delta_{*21} + \Delta'_{*12} M_{*12} \Delta_{*22} + \Delta'_{*22} M_{*21} \Delta_{*12} + \Delta'_{*22} M_{*22} \Delta_{*22} - \bar{\Delta}'_1 \bar{M}_1 \bar{\Delta}_1$$

(B.7) $$\bar{\Delta}_1 = \Delta_{*22} + \bar{M}^{-1}_1 M_{*21} \Delta_{*12}$$

(B.8) $$\bar{M}_1 = M_{*22}.$$

It can be seen that $|B'_{\bar{1}} \; B_{\bar{1}}|$ and $|\bar{S}_1 + (\Delta_{\bar{1}} - \bar{\Delta}_1)' \bar{M}_1 (\Delta_{\bar{1}} - \bar{\Delta}_1)|$ do not depend on δ_1, but they depend on $\delta_{\bar{1}}$ through the free parameters of $B_{\bar{1}}$ and the "column-diagonal" matrix $\Delta_{\bar{1}}$ (selected from Δ) respectively. This justifies their presence in the kernel of the marginal density of $\delta_{\bar{1}}$.

B.II. *COMPUTATION OF THE POSTERIOR MODE OF δ*

The computation of the mode d of (3.1) requires the use of a standard optimization technique, e.g. iteration of the form

(B.9) $$d_{i+1} = d_i - H^{-1}_{d_i} \cdot g_{d_i}$$

or other standard techniques, see e.g. Harvey (1981, Chapter 4).

In (B.9), g_{d_i} is the gradient $\partial \log p(\delta)/\partial \delta$ and H_{d_i} is the Hessian matrix

$\partial^2 \log p(\delta)/\partial\delta\,\partial\delta'$ (or an approximation of it), both evaluated at $\delta = d_i$.

Morales (1971) has derived analytically g_δ and H_δ and shown that computing the posterior mode was formally equivalent to computing a FIML estimate. We give now for ease of reference the formulas of g_δ and H_δ in a compact notation :

(B.10) $$g_\delta = -\tau_* \tilde{b} + (C_\delta^{-1} \square N_*) i_m - (C_\delta^{-1} \square M_*)\,\delta$$

(B.11) $$H_\delta = -\tau_* \tilde{b}\tilde{b}' - (C_\delta^{-1} \square M_*) + \frac{2}{\nu_*} E_\delta' F_\delta E_\delta ,$$

(in formula 3.23 of Morales, $2/\eta_2^2$ should be $4/\eta_2^2$, and in 3.24, $1/\eta_2$ should be $2/\eta_2$, corresponding to $2/\nu_*$ in our notation),

where

$\tilde{b} = \dfrac{\partial \log |B|}{\partial\delta} = (\tilde{b}_1 \dots \tilde{b}_\ell)'$; $\tilde{b}_k = 0$ if k is the position in δ of a predetermined variable; $\tilde{b}_k = -b^{h\ell}$ if k is the position in δ of the coefficient of an endogenous variable appearing in $-B$ as $b_{\ell h}$,

$N_* = M_* \Delta_*$ $(\ell \times m)$

$C_\delta^{-1} = \nu_* Q_*^{-1} = (c^{ij})$, $Q_* = S_* + (\Delta - \Delta_*)' M_* (\Delta - \Delta_*)$,

$F_\delta = [F_{ij}]$ $(m^2 \times m^2)$, with F_{ij} $(m \times m)$ given by

$F_{ij} = c^{ij} C_\delta^{-1}, + c^i c^{j'}$, c^i being the i-th column of C_δ^{-1},

$$E_\delta = \begin{bmatrix} E_1 & 0 & \dots & 0 \\ 0 & E_2 & \dots & 0 \\ \vdots & \vdots & \ddots & \vdots \\ 0 & 0 & \dots & E_m \end{bmatrix}$$ $(m^2 \times \ell)$, E_i' $(\ell_i \times m)$ being the i-th row block of the matrix $N_* - M_* \Delta$

B.III. <u>COMPUTATION OF (3.15)</u>

In order to compute (3.15), we use again the technique of importance sampling, i.e. we approximate the expected value of the right-hand side of (3.15) by a sample average :

(B.12) $$f(\delta) \simeq \frac{1}{L} \sum_{k=1}^{L} \prod_{i=1}^{m} p(\delta_i \mid \phi_{\bar{i}k})$$

where the $\phi_{\bar{i}k}$'s $(i = 1,\ldots,m,\ k = 1,\ldots,L)$ are obtained from a random sample $\phi_1 \ldots \phi_L$ generated from the Student density of ϕ. In practice, one can of course use repeatedly the same sample $\phi_1 \ldots \phi_L$ for the different values of δ at which $f(\delta)$ must be computed by formula (B.12).

Moreover, since random drawings from $h(\phi)$ are anyway needed to generate the N drawings of δ in the two-stage procedure described in subsection III.2.4., one can also use the same sample $\phi_1 \ldots \phi_L$ as conditioning values needed for the second stage of that procedure; if $L < N$, one will use that sample several times, while if $L > N$, one will use a subsample. In this way, many Student drawings are saved, as well as many computations required to generate 1-1 poly-t random drawings, at the cost of additional storage requirements. This applies also when (3.14) is the importance function. Note that in this case, there is no error of approximation in the computation of the value of the importance function. However, (3.14) and (B.12) are essentially equivalent when (3.12) is built as an approximation of (3.11); in (B.12), $w_k = 1/L$ but the values $\phi_{\bar{i}k}$ represent the probabilistic Student structure. For example, in (3.14) there is one and only one value $\phi_{\bar{i}k}$ corresponding to the mode of the Student law, but w_k is the probability of a hypercube centered at the mode; in (B.12), there will be approximately the same proportion w_k of values $\phi_{\bar{i}k}$ located in that hypercube, provided L is large enough. Clearly, in both cases, L has to be sufficiently large : for (3.14), it should be an exponential function of ℓ, say $L = a^{\ell}$ for some a; for (B.12), it should be chosen so as to guarantee a certain relative precision $100 - \bar{\varepsilon}$ with probability $1 - \alpha$, as usual in a Monte Carlo setup - see formula (2.33) solved for N, standing for L, where σ/μ_g represent in this context the variation coefficient of $\prod_{i=1}^{m} p(\delta_i \mid \phi_{\bar{i}})$ with respect to $h(\phi)$; note that this variation coefficient and therefore L vary with the particular value of δ, so that some representative value of L has to be chosen.

APPENDIX C : PLOTS OF POSTERIOR MARGINAL DENSITIES AND OF IMPORTANCE FUNCTIONS

BBM
FIGURE 1-A : BETA1

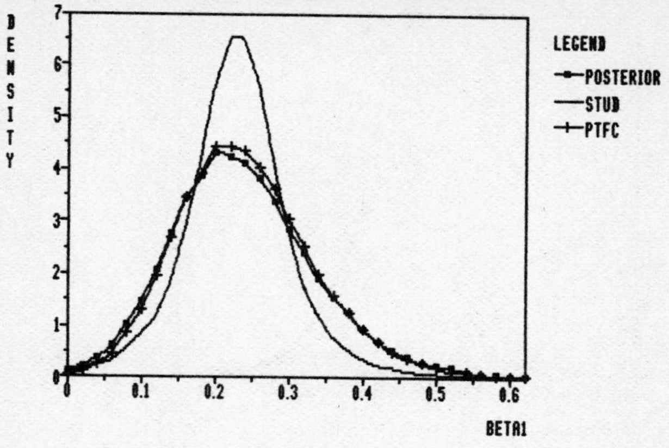

BBM
FIGURE 1-B : BETA1

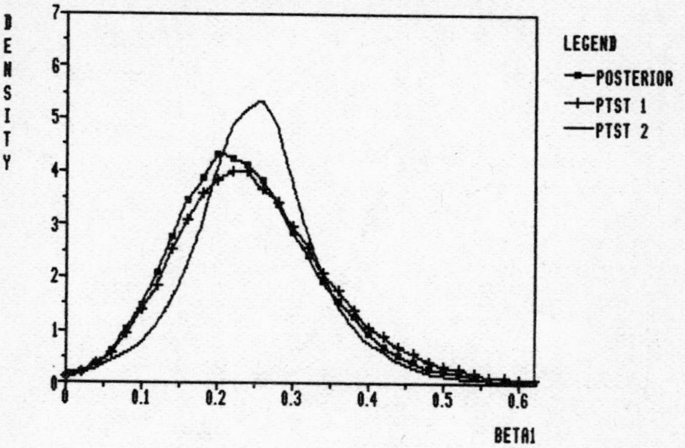

BBM
FIGURE 1-C : BETA1

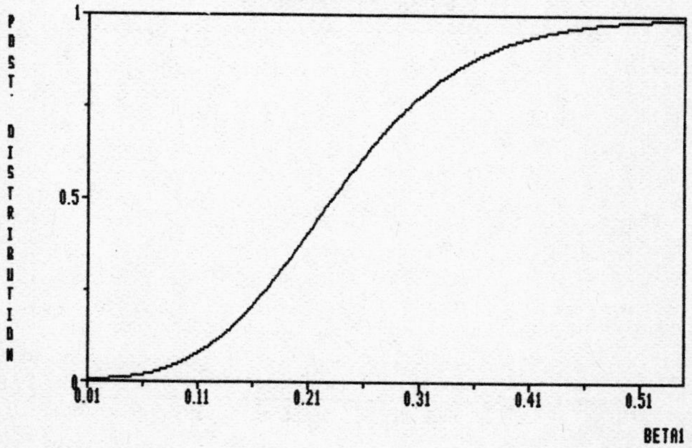

BBM
FIGURE 2-A : GAMMA4

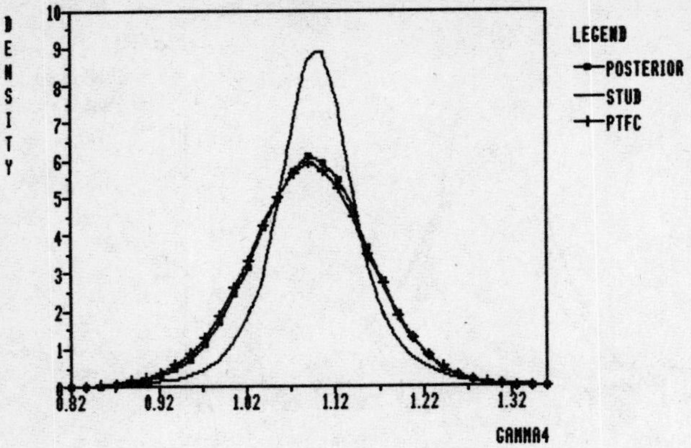

BBM
FIGURE 2-B : GAMMA4

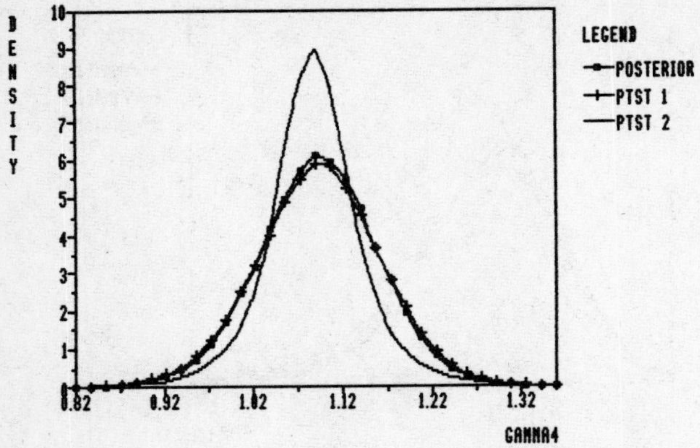

BBM
FIGURE 2-C : GAMMA4

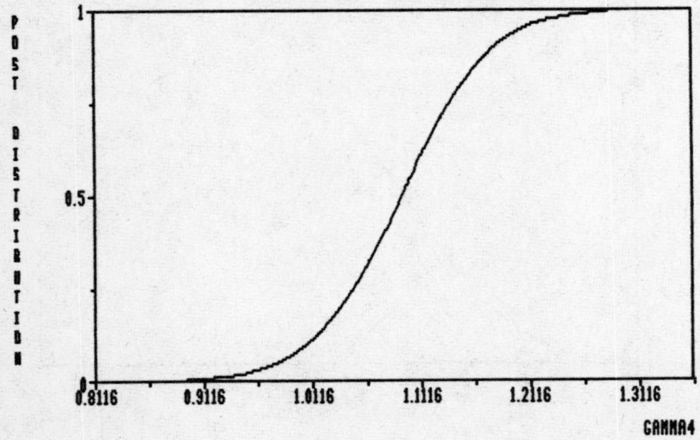

BBM
FIGURE 3-A : BETA2

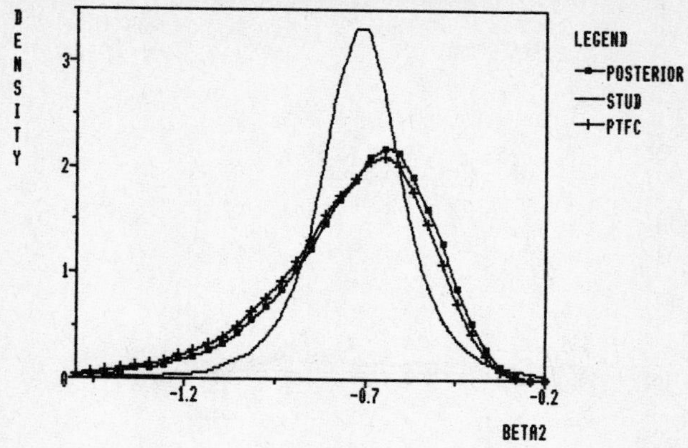

BBM
FIGURE 3-B : BETA2

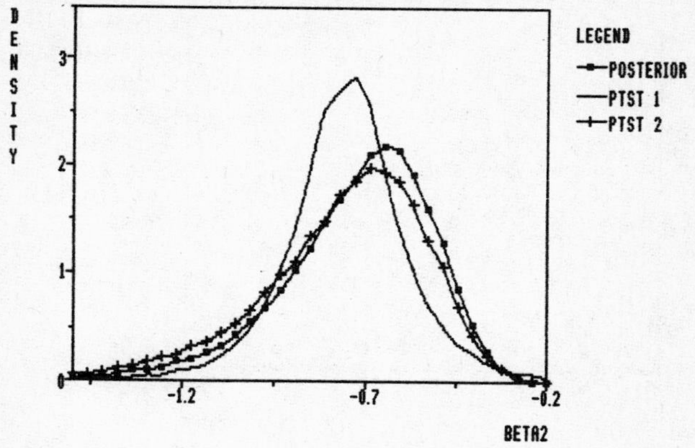

BBM
FIGURE 3-C : BETA2

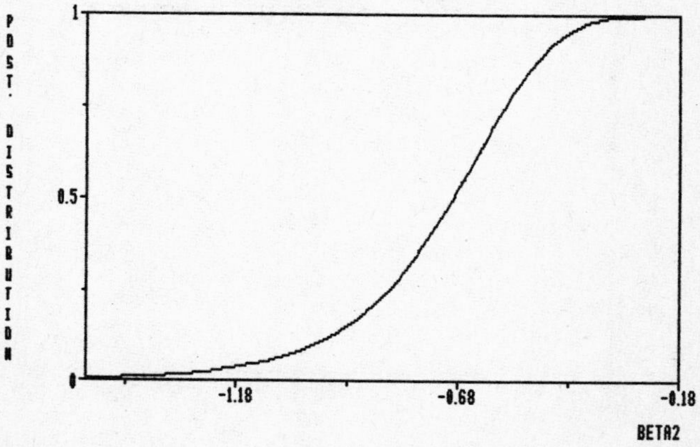

BBM
FIGURE 4-A : GAMMA3

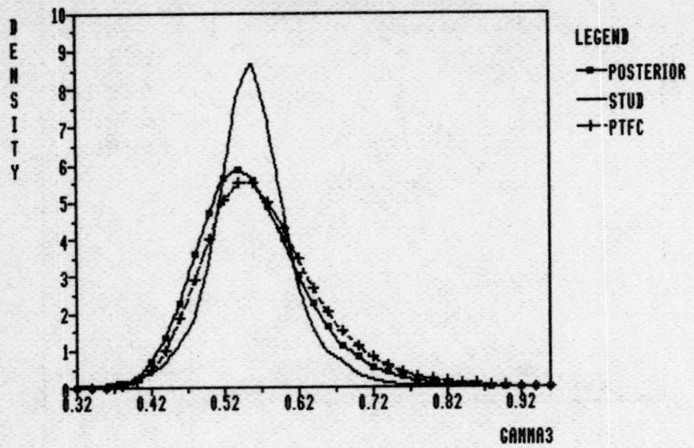

BBM
FIGURE 4-B : GAMMA3

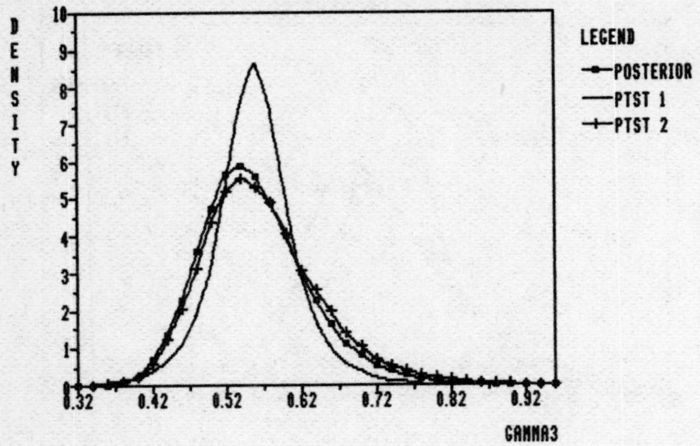

BBM
FIGURE 4-C : GAMMA3

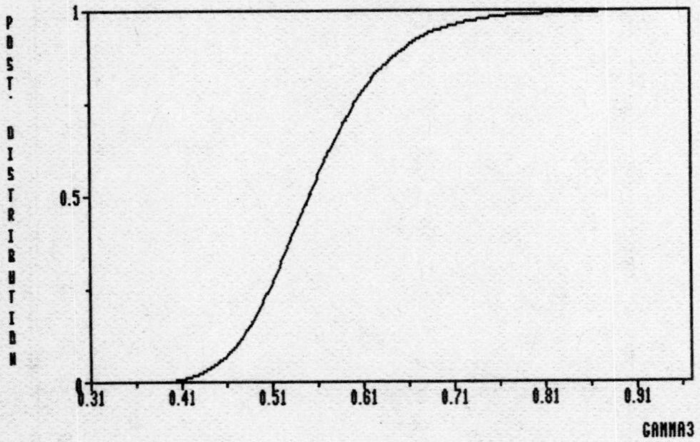

JOHNSTON
FIGURE 5-A : BETA1

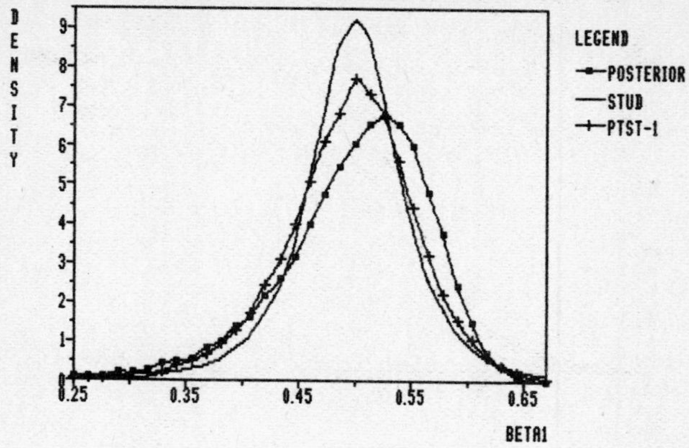

JOHNSTON
FIGURE 5-B : BETA1

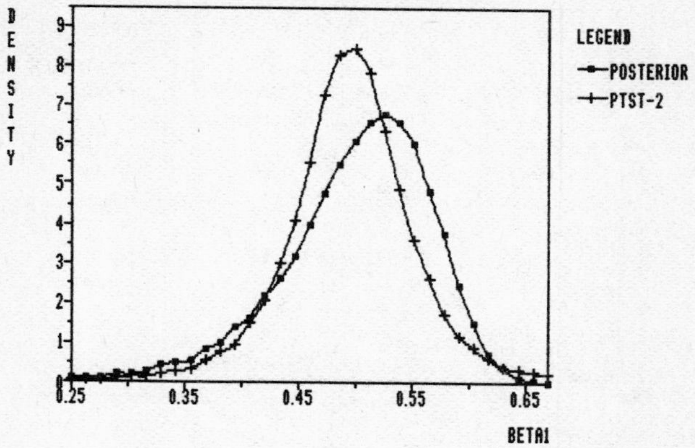

JOHNSTON
FIGURE 5-C : BETA1

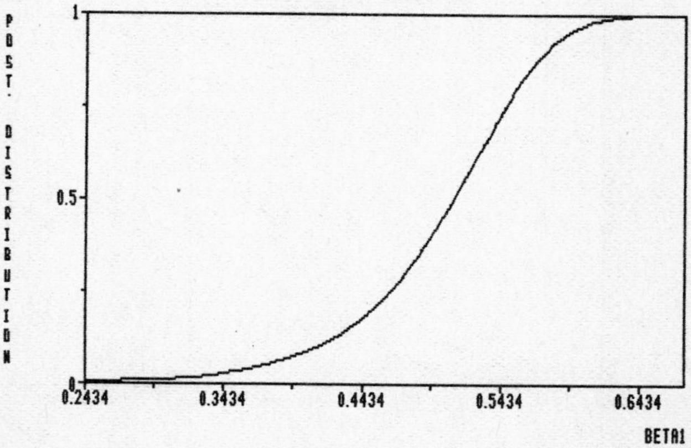

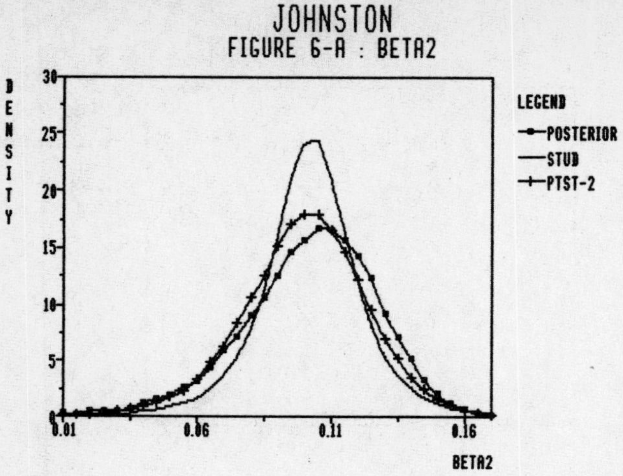
JOHNSTON
FIGURE 6-A : BETA2
DENSITY
30
25
20
15
10
5
0
0.01
0.06
0.11
0.16
BETA2
LEGEND
POSTERIOR
STUD
PTST-2

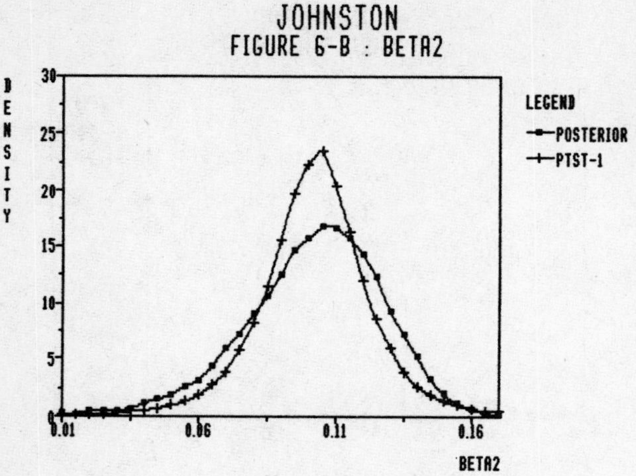
JOHNSTON
FIGURE 6-B : BETA2
DENSITY
30
25
20
15
10
5
0
0.01
0.06
0.11
0.16
BETA2
LEGEND
POSTERIOR
PTST-1

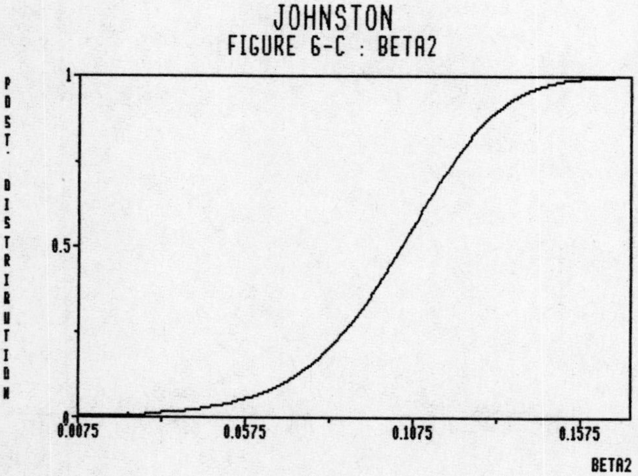
JOHNSTON
FIGURE 6-C : BETA2
POST. DISTRIBUTION
1
0.5
0
0.0075
0.0575
0.1075
0.1575
BETA2

JOHNSTON
FIGURE 7-A : GAMMA2

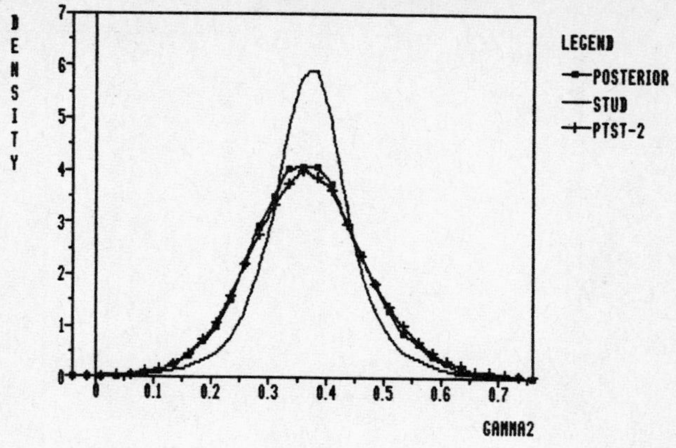

JOHNSTON
FIGURE 7-B : GAMMA2

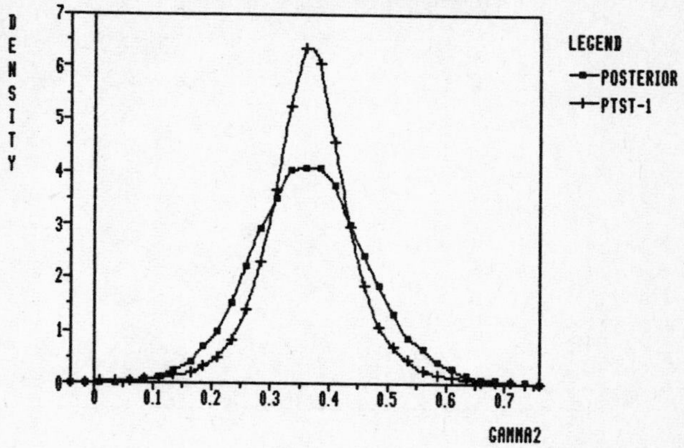

JOHNSTON
FIGURE 7-C : GAMMA2

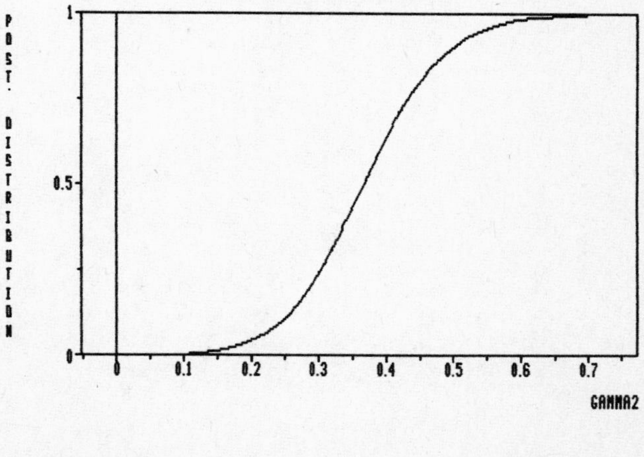

JOHNSTON

FIGURE 8-A : SRM OF Z ON Y

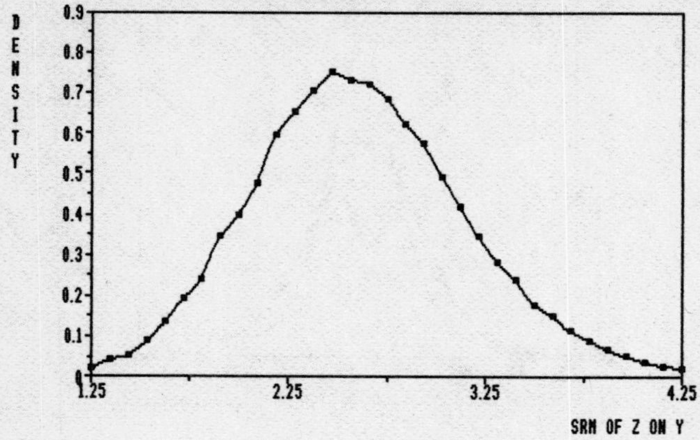

JOHNSTON

FIGURE 8-B : SRM OF Z ON Y

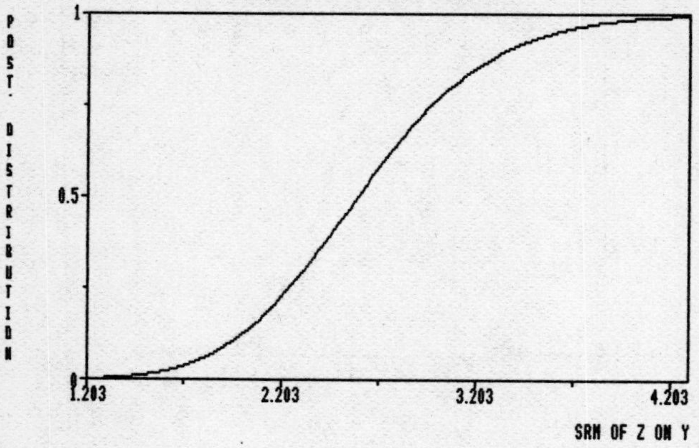

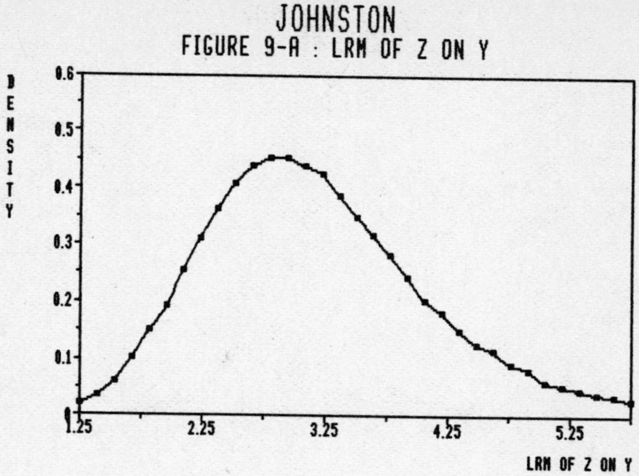
JOHNSTON
FIGURE 9-A : LRM OF Z ON Y
DENSITY
0.6
0.5
0.4
0.3
0.2
0.1
0
1.25
2.25
3.25
4.25
5.25
LRM OF Z ON Y

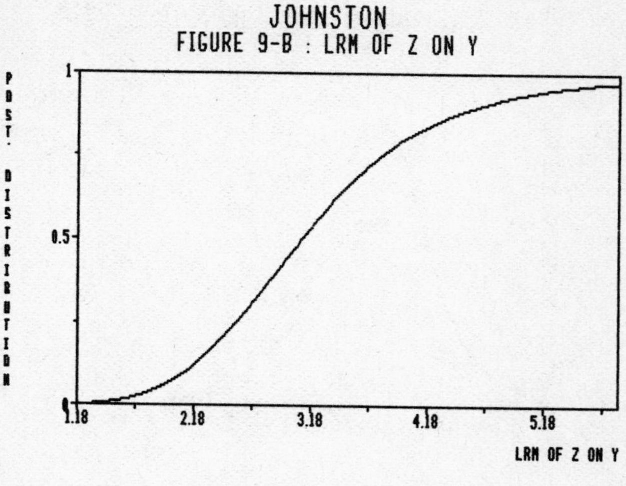
JOHNSTON
FIGURE 9-B : LRM OF Z ON Y
POST. DISTRIBUTION
1
0.5
0
1.18
2.18
3.18
4.18
5.18
LRM OF Z ON Y

KLEIN : NUO = 8
FIGURE 10-A : ALPHA1

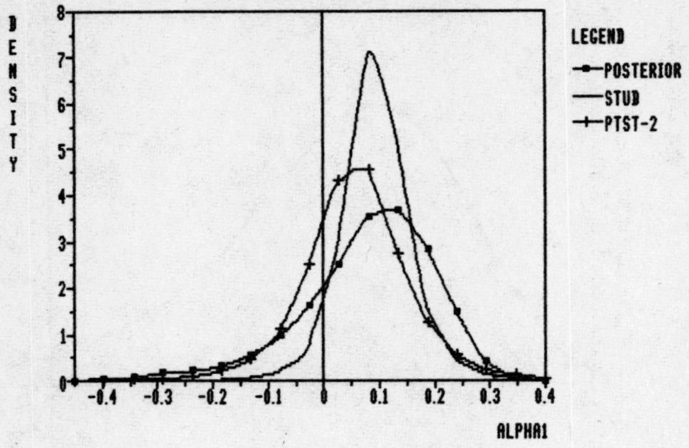

KLEIN : NUO = 8 & TRUNCATED PRIOR
FIGURE 10-B : ALPHA1

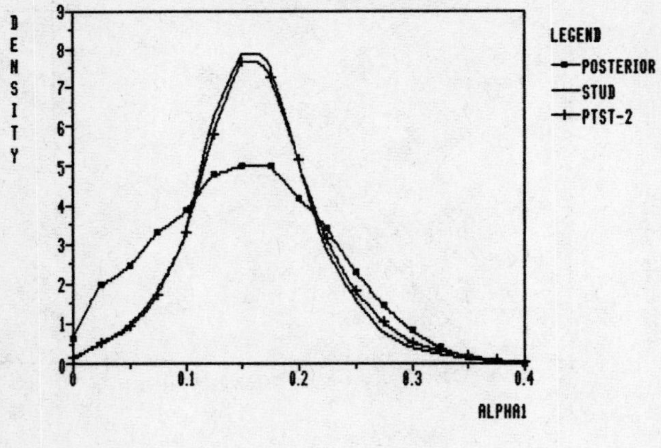

KLEIN : NU0 = 8
FIGURE 11-A : ALPHA2

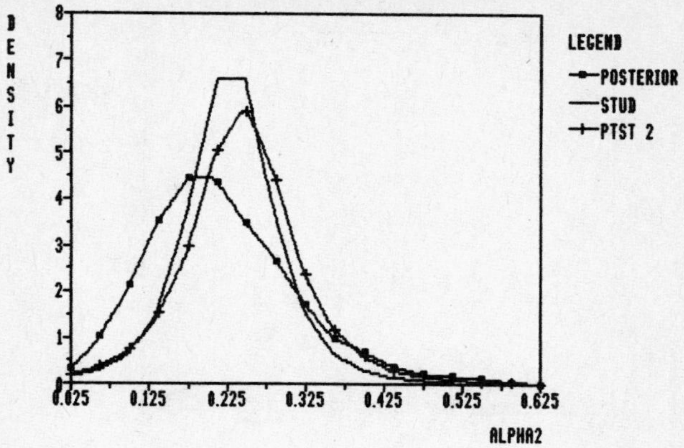

KLEIN : NU0 = 8 & TRUNCATED PRIOR
FIGURE 11-B : ALPHA2

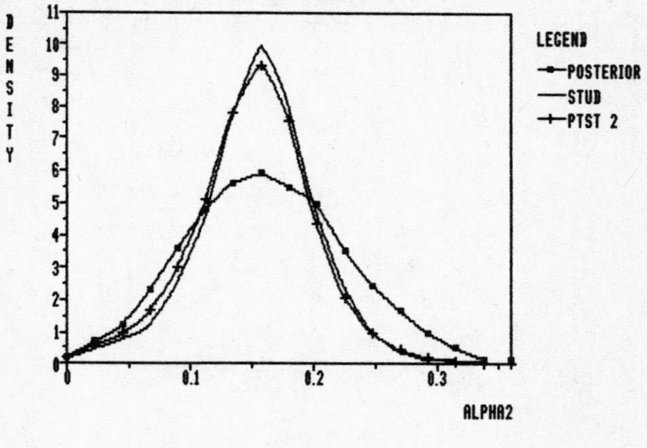

KLEIN : NU0 = 8
FIGURE 12-A : ALPHA3

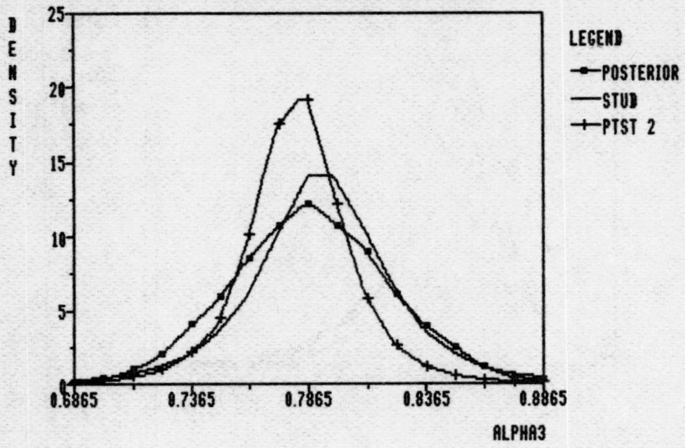

KLEIN : NU0 = 8 & TRUNCATED PRIOR
FIGURE 12-B : ALPHA3

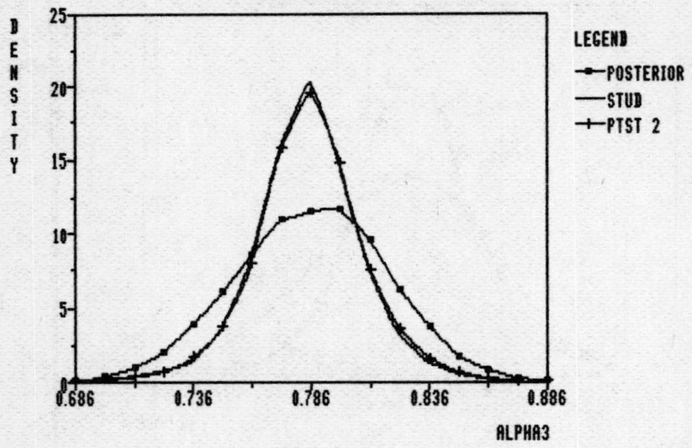

KLEIN : NU0 = 8

FIGURE 13-A : BETA1

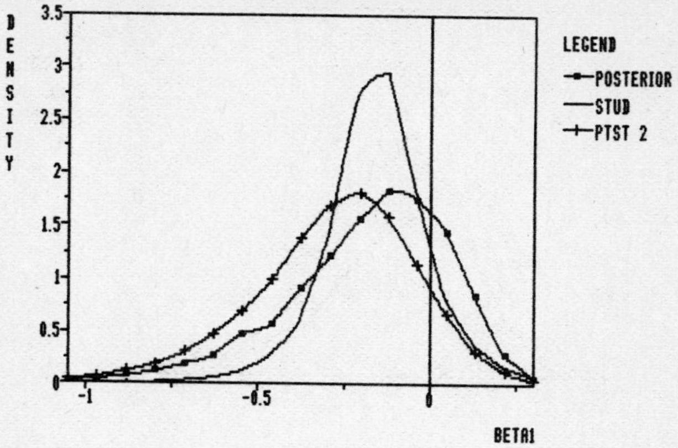

KLEIN : NU0 = 8 & TRUNCATED PRIOR

FIGURE 13-B : BETA1

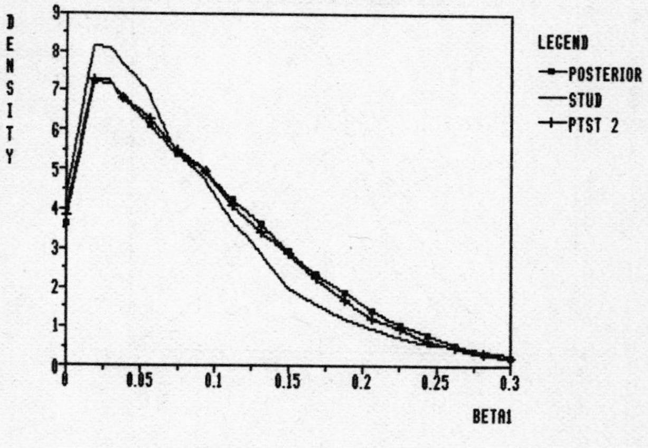

KLEIN : NU0 = 8
FIGURE 14-A : BETA2

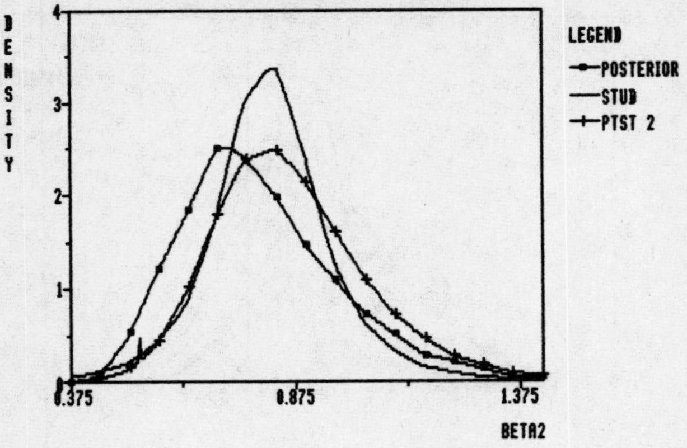

KLEIN : NU0 = 8 & TRUNCATED PRIOR
FIGURE 14-B : BETA2

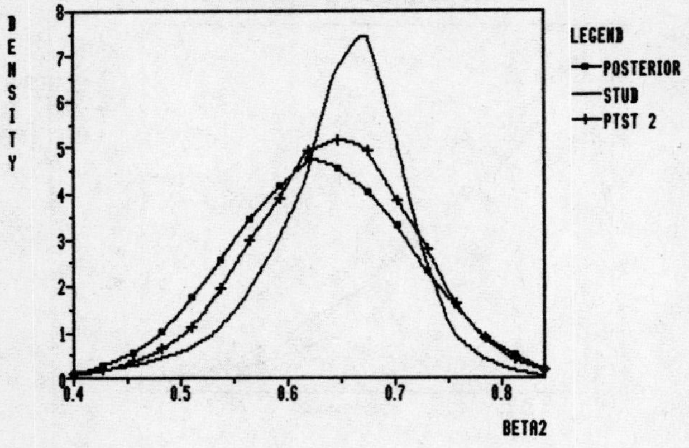

KLEIN : NU0 = 8
FIGURE 15-A : BETA3

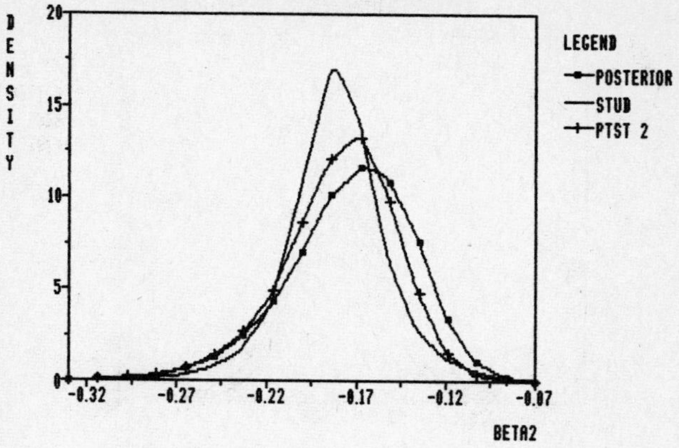

KLEIN : NU0 = 8
FIGURE 15-B : BETA3 & TRUNCATED PRIOR

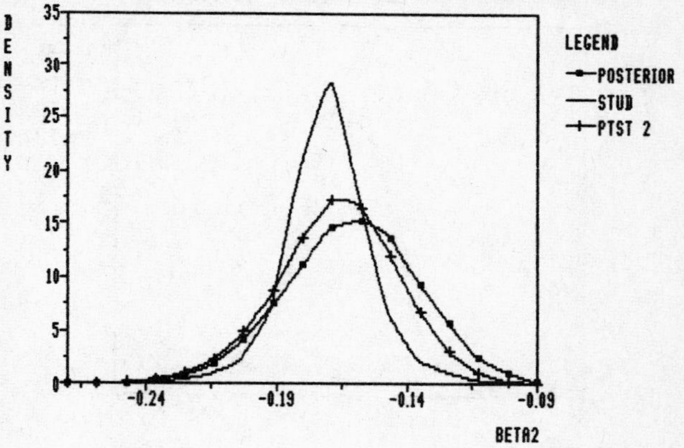

KLEIN : NU0 = 8
FIGURE 16-A : GAMMA1

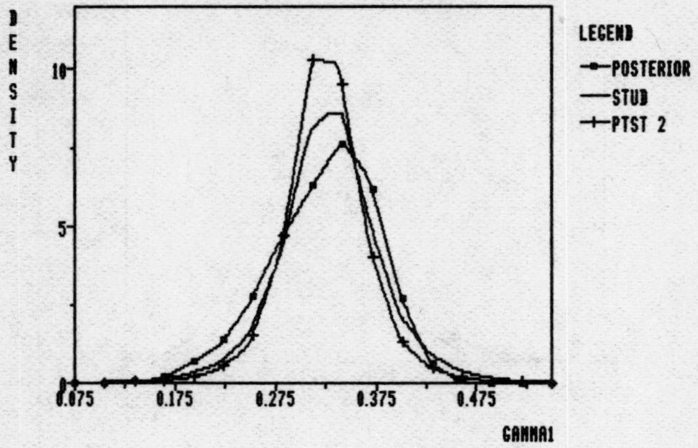

KLEIN : NU0 = 8 & TRUNCATED PRIOR
FIGURE 16-B : GAMMA1

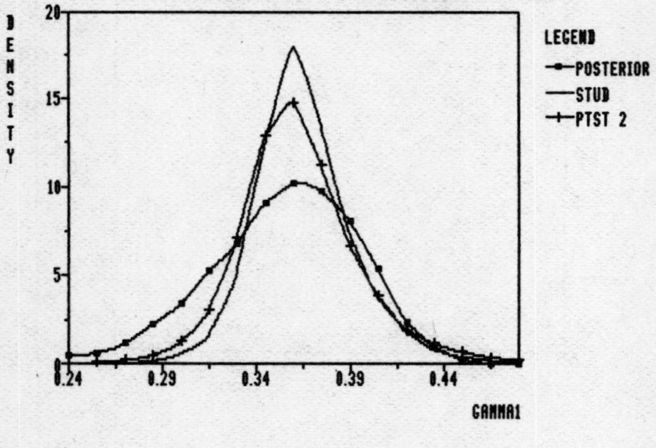

KLEIN : NU0 = 8
FIGURE 17-A : GAMMA2

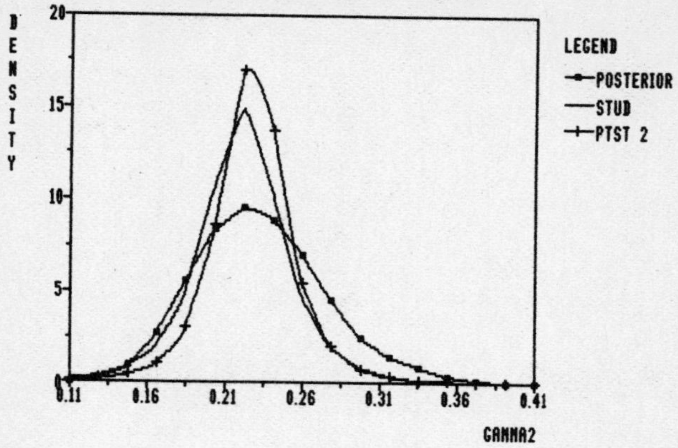

KLEIN : NU0 = 8 & TRUNCATED PRIOR
FIGURE 17-B : GAMMA2

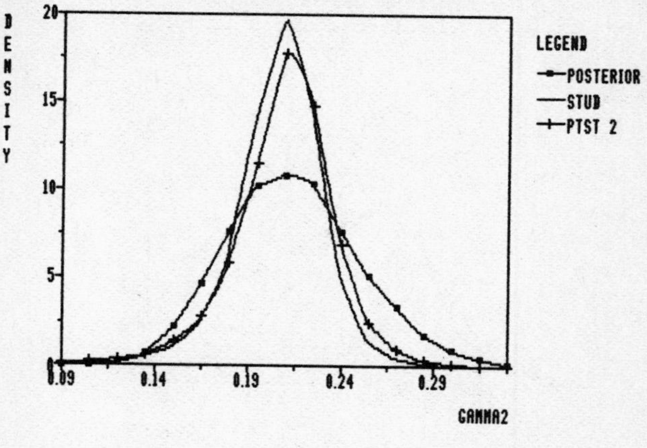

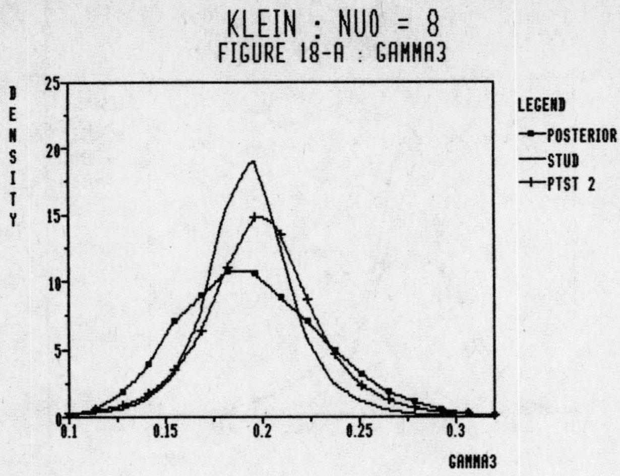
KLEIN : NU0 = 8
FIGURE 18-A : GAMMA3
DENSITY
25
20
15
10
5
0
0.1
0.15
0.2
0.25
0.3
GAMMA3
LEGEND
POSTERIOR
STUD
PTST 2

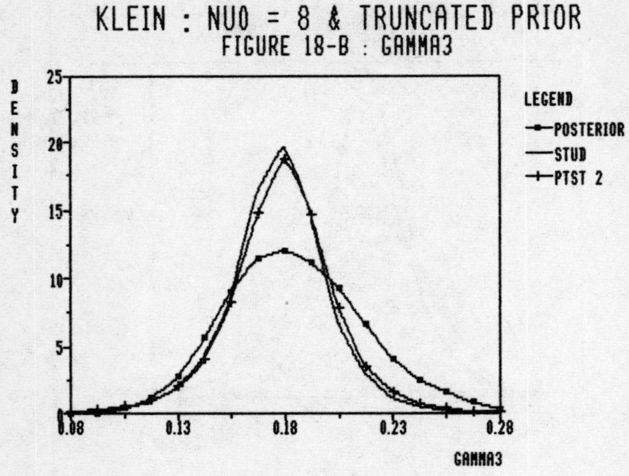
KLEIN : NU0 = 8 & TRUNCATED PRIOR
FIGURE 18-B : GAMMA3
DENSITY
25
20
15
10
5
0
0.08
0.13
0.18
0.23
0.28
GAMMA3
LEGEND
POSTERIOR
STUD
PTST 2

EX

FIGURE 19-A : E(XI;W)

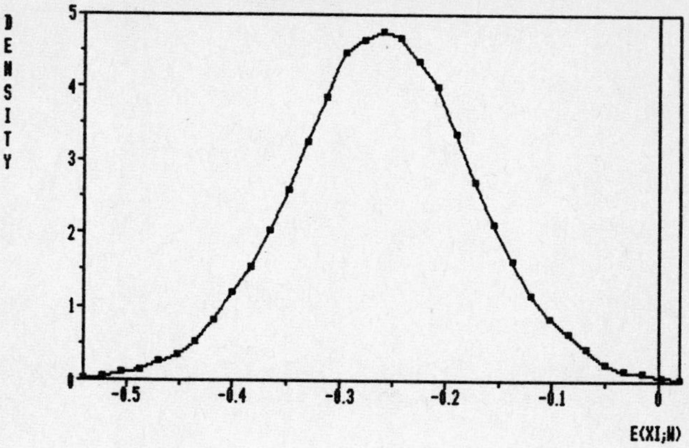

FIGURE 19-B : E(XI;W)

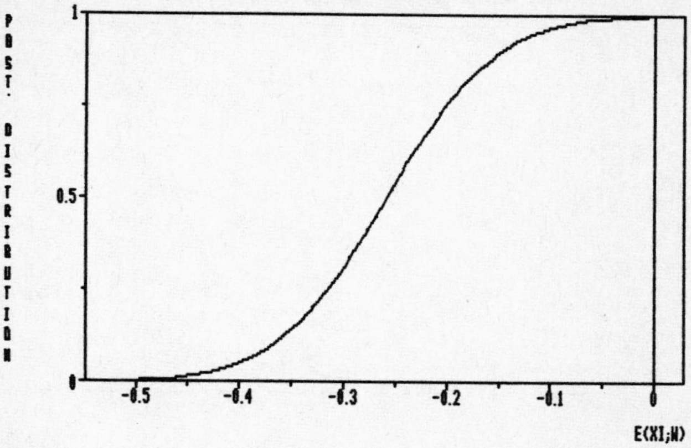

EX

FIGURE 20-A : E(L;W/P), SHORT-RUN

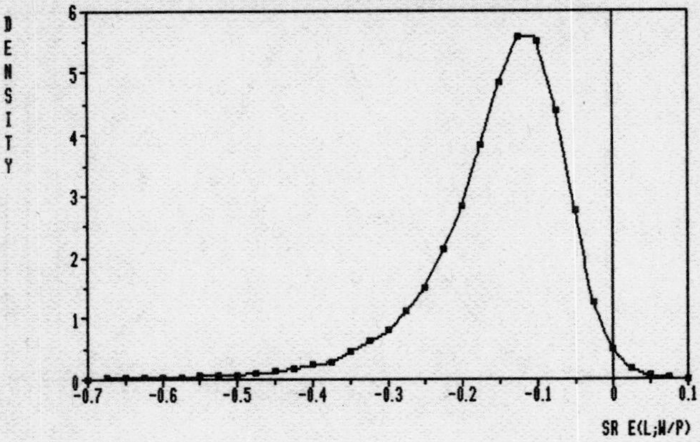

FIGURE 20-B : E(L;W/P), SHORT-RUN

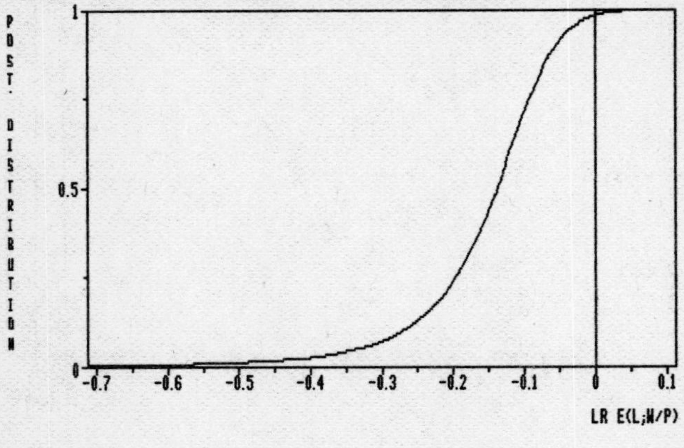

EX

FIGURE 21-A : E(L;W/P), LONG-RUN

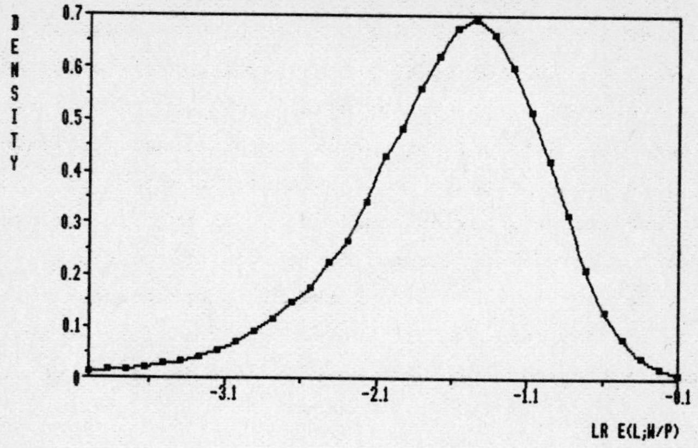

FIGURE 21-B : E(L;W/P), LONG-RUN

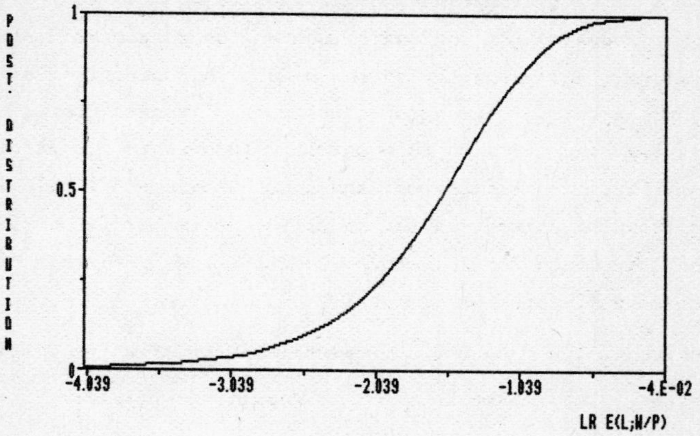

APPENDIX D : THE COMPUTER PROGRAM

We have integrated our programs of Monte Carlo integration of the posterior density of the simultaneous equation model in the Bayesian Regression Program (BRP) developed at CORE and documented in Bauwens et al. (1981). BRP has facilities for entering and transforming data and for matrix computations. It includes also the (conditional) full information analysis of a single equation (FI) and the treatment of the seemingly unrelated regression model (SU), through an evaluation of the first two moments and of the univariate marginal densities of the conditional posterior density $p(\delta_1 | \delta_{\bar{1}})$ defined by (3.3), using the PTD subprograms. These 2 options can now be replaced by the new option MCFI (Monte Carlo Full Information) which implements a more complete treatment of these models along the lines described in this paper.

We review briefly the parts of this new option; each part consists of one or several subroutines. A detailed user's manual is available.

1) MCFI : this program reads the basic inputs (the data matrix, the lists of endogenous and predetermined variables of the model, the specification of each stochastic equation, the nature of the prior density and its parameters, the structure of the B matrix (i.e. the value of this matrix with 0's and 1's at appropriate places, to indicate normalizations, exclusions, and identities), an initial value of the vector δ, and computes the parameters (2.15) of the posterior density, as well as other intermediate results needed for subsequent computations. Several options are then available for treating the posterior density

- computation of the mode and the associated covariance matrix (MAXLP);
- evaluation of the conditional poly-t densities $p(\delta_i | \delta_{\bar{i}})$, $i = 1,\dots,m$ (CPTD);
- conditional bivariate contour plots (PCNTR);
- Monte Carlo integration (MCI).

2) MAXLP : this part organizes the computation of the mode of the log posterior and of (minus) the inverse of the Hessian at the mode. The method of maximisation used is a quasi-Newton (or variable metric) method which is a variant of the Davidon-Fletcher-Powell (DFP) algorithm. Use is made of the GQOPT implementation of DFP (of S. Goldfeld and R. Quandt). The gradient is computed by formula (B.10) but the (inverse of the) Hessian is not evaluated explixitly : it is replaced by a PDS matrix which is updated at each iteration of (B.9) and converges to the inverse of the Hessian. We intend to introduce the computation of (B.11) once the mode is obtained.

3) <u>CPTD</u> : this part evaluates all the conditional poly-t densities $p(\delta_i | \delta_{\bar{i}})$ through the PTD subprograms of BRP. The conditioning values are taken either from the initial value of δ given by the user, or from the posterior mode if it has been computed. The output for each poly-t is the same as described above for the former options FI or SU.

4) <u>PCNTR</u> : for any pair of coefficients of the model, it is possible to obtain a contour plot of their conditional density; the conditions are fixed as in CPTD.

5) <u>MCI</u> : this group of subroutines organizes the Monte Carlo integration of the posterior density. The user selects the importance function through one of the keywords STUD, PTFC, PTDC or PTST. For each of these, there are subprograms for

- preparing the drawing and the computation of the importance function,
- drawing from the importance function and cumulating the weighted sums (2.29),
- computing the value of the ratio of the kernel of the posterior to that of the importance function.

Another subset of programs reads what results are asked for, and prints the final results. Many options are available, among which :

- results for the structural coefficients (by default) and on some or all reduced form coefficients (the matrix Π is computed automatically at each drawing, on request, if the structure of the matrix Γ is given); it is also possible to add FORTRAN statements, that compute other functions of δ, in the subroutine where the weighted sums are cumulated. The results are, by default, the expected values, the skewness coefficients and the covariance matrix of all the functions of δ for which results are asked for; optionally : univariate marginal densities (a plot of 33 points at most) and fractiles, contour plots of bivariate marginal densities, for which the user must give the ranges of each coefficient;

- the same results for the importance function;

- the expected values of the covariance matrices of the disturbances of the structural and reduced forms (Σ and Ω).

The values of the cumulated sums are always saved at the end of a run so that the user can resume execution from that point. It is also possible in any run to save these values regularly in an auxiliary file; an auxiliary program is available for computing the corresponding results.

The program is written in FORTRAN 77 and is implemented on a Data General MV8000 computer.

FOOTNOTES

[1]Other normalisation rules could be considered at the cost of more complicated notation.

[2]DR (section 6.5) treat implicitly the case where Δ_0 is "column-diagonal".

[3]Except if $\tau_0 = 0$ or $|B| = 1$, $\nu_0 > m-1$ and M_0 is PDS, and even if Δ_0 is not "column-diagonal", as can be easily seen from (2.7), see lemma 2.1 of Morales (1971) for the case where Δ_0 is "column-diagonal"; if the conditions (i) - (v), stated after (2.3), are verified the integrating constant of the density $p(\Sigma)$ is given by

$$E\left[\prod_{i=1}^{m} (\sigma^{ii})^{-\frac{1}{2}\ell_i}\right] \cdot K_m(\nu_0, S_0)$$

where $K_m(\nu_0, S_0)$ is the integrating constant of a Wishart distribution for Σ^{-1}, with parameters ν_0 and S_0, and the expected value is taken with respect to this Wishart distribution; see Johnson and Kotz (1972, p.164) for the computation of such an expected value.

[4]It is implicit in this notation that if w is present in the subscript of μ, the expected value is computed using the density $f(\theta)$; if w is not present, as in (2.26), the density used is $p(\theta)$.

[5]This can be diagnosed only in the case where another importance function can be found, that is a good approximation.

[6]From here on, $p(\delta_i|\delta_{\bar{i}})$, $p(\delta)$, etc. stand for $p(\delta_i|\delta_{\bar{i}}, X)$, $p(\delta|X)$, etc.

[7]Formula (6.1.4) of Hammersley and Handscomb becomes in our notation

$$w(\delta,\phi) = p(\delta)\, g(\delta,\phi)\, J(\delta,\phi)/G(\delta) \prod_{i=1}^{m} p(\delta_i|\phi_{\bar{i}})\, k(\phi);$$

we have a particular application since we set

$$g(\delta,\phi) = \prod_{i=1}^{m} p(\delta_i|\phi_{\bar{i}})\, h(\phi),$$

so that $G(\delta) = \int g(\delta,\phi)\, d\phi$ is nothing else but $f(\delta)$ as given by (3.13); we have an extreme application since the space of ϕ is the same as the space of δ, so that $J(\delta,\phi) = 1$, and $w(\delta,\phi) = w(\delta) = p(\delta)/f(\delta)$.

[8]e.g. we could define the $\tilde{E}(\ldots)$ functions of (3.29) by fixing once for all the elements depending on γ, but by renewing at each drawing the elements depending on β (using the drawn value of β).

[9]We had to scale the data by multiplying y_1 and y_2 by .1, z_3 by .0001 and z_4 by 10, so that all data are more or less of the order of unity, in order to avoid overflow in computing the value of the posterior density. Consequently, for com-

parison with the results reported by Richard, the dimension of the parameters as reported here is to be adjusted as follows : β_1 and β_2 : unchanged; γ_3 : $\times$ .001; γ_4 : $\times$ 100; Σ : $\times$ 100.

[10]i.e. the simple correlation coefficients computed from the posterior variance - covariance matrix.

[11]We do not assert that this is true in general.

[12]Except by a rejection method (see Appendix A), which would not be very convenient in this context.

[13]Drèze and Modigliani used another version of the model, where $\beta_6 = 1-\alpha$; this restriction was imposed to "solve" a problem of collinearity of the data.

[14]With a diffuse prior, $S_* = S = Y'HY$, where $H = I - \Xi(\Xi'\Xi)^{-1}\Xi'$; the rank of S is not greater than the minimum of the rank of Y and of H, i.e. m and $T-\ell$; if $m = 6$, $T = 22$, $\ell = 18$, $r(S) \leqslant 4$; if $m = 5$, $T = 22$, $\ell = 15$, $r(S) \leqslant 5$ but with the data we had, if we drop the equation of the United Kingdom, it turns out that S is not of full rank; therefore we drop also the equation of Italy.

[15]This number depends on the number of parameters; in our experiments, 1000 drawings have always been sufficient.

[16]For the seemingly unrelated regression model, 2-0 poly-t densities replace 1-0 (Student) ones, and it is not much more costly to generate random drawings from the former than from the latter.

[17]See VDK (1982 a) for an application to the Klein-Goldberger model.

REFERENCES

BARTEN, A.P., d'ALCANTARA, G., CARRIN, G., (1976), "COMET, a Medium-term Macroeconomic Model for the European Economic Community", *European Economic Review, 7*, 63-115.

BAUWENS, L., BULTEAU, J.P., GILLE, P., LONGREE, L., LUBRANO, M., TOMPA, H., (1981), "Bayesian Regression Program (BRP) User's Manual", Université Catholique de Louvain, CORE Computing Report 81-A-01.

BAUWENS, L., d'ALCANTARA, G., (1981), "An Export Model for the Belgian Industry", Université Catholique de Louvain, CORE Discussion Paper 8105 (forthcoming in *European Economic Review*).

BAUWENS, L., RICHARD, J.-F., (1982), "A Poly-t Random Variable Generator, with Application to Monte Carlo Integration", Université Catholique de Louvain, CORE Discussion Paper 8214.

CHOW, G.C., (1973), "Multiperiod Predictions from Stochastic Difference Equations by Bayesian Methods", *Econometrica, 41*, 109-118.

CRAMER, H., (1946), *Mathematical Methods of Statistics*, Princeton, Princeton University Press.

DAGENAIS, M., (1978), "The Computation of FIML Estimates as Iterative Generalized Least Squares Estimates in Linear and Nonlinear Simultaneous Equations Models", *Econometrica, 46*, 1351-1362.

DAVIS, P.J., RABINOWITZ, P., (1975), *Methods of Numerical Integration*, New York, Academic Press.

DREZE, J.H., MORALES, J.A., (1976), "Bayesian Full Information Analysis of Simultaneous Equations", *Journal of the American Statistical Association, 71*, 919-923.

DREZE, J.H., (1976), "Bayesian Limited Information Analysis of the Simultaneous Equations Model", *Econometrica, 44*, 1045-1075.

DREZE, J.H., MODIGLIANI, F., (1982), "The Trade-off between Real Wages and Employment in an Open Economy (Belgium)", *European Economic Review, 15*, 1-40.

DREZE, J.H., RICHARD, J.-F., (1983), "Bayesian Analysis of Simultaneous Equation Systems", in : Z. Griliches and M.D. Intriligator (eds.), *Handbook of Econometrics*, Vol. I, Part 3, Chapter 9, Amsterdam, North Holland Publishing Co.

ENGLE, R.F., HENDRY, D.F., RICHARD, J.-F., (1983), "Exogeneity", *Econometrica, 51*, 277-304.

FLORENS, J.-P., (1977), "Approximate Posterior Distributions in Two Non Linear Models", Paper presented at the European Meeting of the Econometric Society in Vienna.

GOVAERTS, B., (1983), "Intégration d'une distribution normale ou Student tronquée à un polyèdre", Mimeo, Université Catholique de Louvain, Faculté des Sciences.

HAMMERSLEY, J.M., HANDSCOMB, D.C.,(1979), *Monte Carlo Methods*, London, Chapman and Hall.

HARVEY, A.C., (1981), *The Econometric Analysis of Time Series*, Oxford, Philip Allan.

JOHNSON, N.L., KOTZ, S., (1972), *Distributions in Statistics : Continuous Multivariate Distributions*, New-York, Wiley.

JOHNSTON, J., (1963), *Econometric Methods*, 1st edition, New-York, McGraw-Hill.

KENDALL, M.G., STUART, A., (1963), *The Advanced Theory of Statistics*, Vol.I, London, Charles Griffin & Cy.

KENNEDY, W.S., GENTLE, J.E., (1980), *Statistical Computing*, New-York and Basel, Marcel Dekker.

KINDERMAN, A.J., MONAHAN, J.F., (1980), "New Methods for Generating Student's and Gamma Variables", *Computing, 25*, 369-377.

KLEIN, L.R., (1950), *Economic Fluctuations in the United States*, New-York, Wiley.

KLOEK, T., VAN DIJK, H.K., (1978), "Bayesian Estimates of Equation System Parameters : An Application of Integration by Monte Carlo", *Econometrica, 46*, 1-19.

KNUTH, D.E., (1971), *The Art of Computer Programming*, Vol. 2, Reading (Mass.), Addison-Wesley.

LUBRANO, M., (1983), "Bayesian Analysis of Switching Regression Models", Université Catholique de Louvain, CORE Discussion Paper 8332.

MALINVAUD, E., (1978), *Méthodes statistiques de l'économétrie*, Paris, Dunod.

MORALES, J.A., (1971), *Bayesian Full Information Structural Analysis*, Berlin, Springer-Verlag.

ODELL, P.L., FEIVESON, A.H., (1966), "A Numerical Procedure to Generate a Sample Covariance Matrix", *Journal of the American Statistical Association, 61*, 199-203.

RICHARD, J.-F., (1973), *Posterior and Predictive Densities for Simultaneous Equation Models*, Berlin, Springer-Verlag.

RICHARD, J.-F., (1979), "Exogeneity, Inference and Prediction in So-called Incomplete Dynamic Simultaneous Equation Models", Université Catholique de Louvain, CORE Discussion Paper 7922.

RICHARD, J.-F., TOMPA, H., (1980), "On the Evaluation of Poly-t Density Functions", *Journal of Econometrics, 12*, 335-351.

ROTHENBERG, T.J., LEENDERS, C.T., (1964), "Efficient Estimation of Simultaneous Equation Systems", *Econometrica, 32*, 57-76.

SPANOS,A., (1982), "Equilibrium, Identities, Behavioural and Measurement Equations in Economic Modelling", University of London, Birkbeck College, Dept. of Economics, Discussion Paper n° 110.

THEIL, H., (1971), *Principles of Econometrics*, New-york, Wiley.

VAN DIJK, H.K., KLOEK, T., (1977), "Predictive Moments of Simultaneous Econometric Models : A Bayesian Approach", in : A. Aykac and C. Brumat (eds.), *New Developments in the Applications of Bayesian Methods*, Amsterdam, North Holland Publishing Co.

VAN DIJK, H.K., KLOEK, T., (1980), "Further Experience in Bayesian Analysis Using Monte Carlo Integration", *Journal of Econometrics, 14*, 307-328.

VAN DIJK, H.K., KLOEK, T., (1981), "Some Alternatives for Simple Importance Sampling in Monte Carlo Integration", Erasmus Universiteit Rotterdam, Working Paper.

VAN DIJK, H.K., KLOEK, T., (1982 a), "Posterior Moments of the Klein-Goldberger Model", Report 8221/E, Econometric Institute, Erasmus Universiteit Rotterdam, to appear in : P.K. Goel and A. Zellner (eds.), *Bayesian Inference and Decision Techniques with Applications*, Amsterdam, North-Holland Publishing Co.

VAN DIJK, H.K., KLOEK, T., (1982 b), "Addendum to : Some Alternatives for Simple Importance Sampling in Monte Carlo Integration", Erasmus Universiteit Rotterdam, Working Paper.

ZELLNER, A., (1971), *An Introduction to Bayesian Inference in Econometrics*, New-York, Wiley.

Vol. 157: Optimization and Operations Research. Proceedings 1977. Edited by R. Henn, B. Korte, and W. Oettli. VI, 270 pages. 1978.

Vol. 158: L. J. Cherene, Set Valued Dynamical Systems and Economic Flow. VIII, 83 pages. 1978.

Vol. 159: Some Aspects of the Foundations of General Equilibrium Theory: The Posthumous Papers of Peter J. Kalman. Edited by J. Green. VI, 167 pages. 1978.

Vol. 160: Integer Programming and Related Areas. A Classified Bibliography. Edited by D. Hausmann. XIV, 314 pages. 1978.

Vol. 161: M. J. Beckmann, Rank in Organizations. VIII, 164 pages. 1978.

Vol. 162: Recent Developments in Variable Structure Systems, Economics and Biology. Proceedings 1977. Edited by R. R. Mohler and A. Ruberti. VI, 326 pages. 1978.

Vol. 163: G. Fandel, Optimale Entscheidungen in Organisationen. VI, 143 Seiten. 1979.

Vol. 164: C. L. Hwang and A. S. M. Masud, Multiple Objective Decision Making – Methods and Applications. A State-of-the-Art Survey. XII, 351 pages. 1979.

Vol. 165: A. Maravall, Identification in Dynamic Shock-Error Models. VIII, 158 pages. 1979.

Vol. 166: R. Cuninghame-Green, Minimax Algebra. XI, 258 pages. 1979.

Vol. 167: M. Faber, Introduction to Modern Austrian Capital Theory. X, 196 pages. 1979.

Vol. 168: Convex Analysis and Mathematical Economics. Proceedings 1978. Edited by J. Kriens. V, 136 pages. 1979.

Vol. 169: A. Rapoport et al., Coalition Formation by Sophisticated Players. VII, 170 pages. 1979.

Vol. 170: A. E. Roth, Axiomatic Models of Bargaining. V, 121 pages. 1979.

Vol. 171: G. F. Newell, Approximate Behavior of Tandem Queues. XI, 410 pages. 1979.

Vol. 172: K. Neumann and U. Steinhardt, GERT Networks and the Time-Oriented Evaluation of Projects. 268 pages. 1979.

Vol. 173: S. Erlander, Optimal Spatial Interaction and the Gravity Model. VII, 107 pages. 1980.

Vol. 174: Extremal Methods and Systems Analysis. Edited by A. V. Fiacco and K. O. Kortanek. XI, 545 pages. 1980.

Vol. 175: S. K. Srinivasan and R. Subramanian, Probabilistic Analysis of Redundant Systems. VII, 356 pages. 1980.

Vol. 176: R. Färe, Laws of Diminishing Returns. VIII, 97 pages. 1980.

Vol. 177: Multiple Criteria Decision Making-Theory and Application. Proceedings, 1979. Edited by G. Fandel and T. Gal. XVI, 570 pages. 1980.

Vol. 178: M. N. Bhattacharyya, Comparison of Box-Jenkins and Bonn Monetary Model Prediction Performance. VII, 146 pages. 1980.

Vol. 179: Recent Results in Stochastic Programming. Proceedings, 1979. Edited by P. Kall and A. Prékopa. IX, 237 pages. 1980.

Vol. 180: J. F. Brotchie, J. W. Dickey and R. Sharpe, TOPAZ – General Planning Technique and its Applications at the Regional, Urban, and Facility Planning Levels. VII, 356 pages. 1980.

Vol. 181: H. D. Sherali and C. M. Shetty, Optimization with Disjunctive Constraints. VIII, 156 pages. 1980.

Vol. 182: J. Wolters, Stochastic Dynamic Properties of Linear Econometric Models. VIII, 154 pages. 1980.

Vol. 183: K. Schittkowski, Nonlinear Programming Codes. VIII, 242 pages. 1980.

Vol. 184: R. E. Burkard and U. Derigs, Assignment and Matching Problems: Solution Methods with FORTRAN-Programs. VIII, 148 pages. 1980.

Vol. 185: C. C. von Weizsäcker, Barriers to Entry. VI, 220 pages. 1980.

Vol. 186: Ch.-L. Hwang and K. Yoon, Multiple Attribute Decision Making – Methods and Applications. A State-of-the-Art-Survey. XI, 259 pages. 1981.

Vol. 187: W. Hock, K. Schittkowski, Test Examples for Nonlinear Programming Codes. V. 178 pages. 1981.

Vol. 188: D. Bös, Economic Theory of Public Enterprise. VII, 142 pages. 1981.

Vol. 189: A. P. Lüthi, Messung wirtschaftlicher Ungleichheit. IX, 287 pages. 1981.

Vol. 190: J. N. Morse, Organizations: Multiple Agents with Multiple Criteria. Proceedings, 1980. VI, 509 pages. 1981.

Vol. 191: H. R. Sneessens, Theory and Estimation of Macroeconomic Rationing Models. VII, 138 pages. 1981.

Vol. 192: H. J. Bierens: Robust Methods and Asymptotic Theory in Nonlinear Econometrics. IX, 198 pages. 1981.

Vol. 193: J. K. Sengupta, Optimal Decisions under Uncertainty. VII, 156 pages. 1981.

Vol. 194: R. W. Shephard, Cost and Production Functions. XI, 104 pages. 1981.

Vol. 195: H. W. Ursprung, Die elementare Katastrophentheorie. Eine Darstellung aus der Sicht der Ökonomie. VII, 332 pages. 1982.

Vol. 196: M. Nermuth, Information Structures in Economics. VIII, 236 pages. 1982.

Vol. 197: Integer Programming and Related Areas. A Classified Bibliography. 1978 – 1981. Edited by R. von Randow. XIV, 338 pages. 1982.

Vol. 198: P. Zweifel, Ein ökonomisches Modell des Arztverhaltens. XIX, 392 Seiten. 1982.

Vol. 199: Evaluating Mathematical Programming Techniques. Proceedings, 1981. Edited by J.M. Mulvey. XI, 379 pages. 1982.

Vol. 200: The Resource Sector in an Open Economy. Edited by H. Siebert. IX, 161 pages. 1984.

Vol. 201: P. M. C. de Boer, Price Effects in Input-Output-Relations: A Theoretical and Empirical Study for the Netherlands 1949–1967. X, 140 pages. 1982.

Vol. 202: U. Witt, J. Perske, SMS – A Program Package for Simulation and Gaming of Stochastic Market Processes and Learning Behavior. VII, 266 pages. 1982.

Vol. 203: Compilation of Input-Output Tables. Proceedings, 1981. Edited by J. V. Skolka. VII, 307 pages. 1982.

Vol. 204: K. C. Mosler, Entscheidungsregeln bei Risiko: Multivariate stochastische Dominanz. VII, 172 Seiten. 1982.

Vol. 205: R. Ramanathan, Introduction to the Theory of Economic Growth. IX, 347 pages. 1982.

Vol. 206: M. H. Karwan, V. Lotfi, J. Telgen, and S. Zionts, Redundancy in Mathematical Programming. VII, 286 pages. 1983.

Vol. 207: Y. Fujimori, Modern Analysis of Value Theory. X, 165 pages. 1982.

Vol. 208: Econometric Decision Models. Proceedings, 1981. Edited by J. Gruber. VI, 364 pages. 1983.

Vol. 209: Essays and Surveys on Multiple Criteria Decision Making. Proceedings, 1982. Edited by P. Hansen. VII, 441 pages. 1983.

Vol. 210: Technology, Organization and Economic Structure. Edited by R. Sato and M. J. Beckmann. VIII, 195 pages. 1983.

Vol. 211: P. van den Heuvel, The Stability of a Macroeconomic System with Quantity Constraints. VII, 169 pages. 1983.

Vol. 212: R. Sato and T. Nôno, Invariance Principles and the Structure of Technology. V, 94 pages. 1983.

Vol. 213: Aspiration Levels in Bargaining and Economic Decision Making. Proceedings, 1982. Edited by R. Tietz. VIII, 406 pages. 1983.

Vol. 214: M. Faber, H. Niemes und G. Stephan, Entropie, Umweltschutz und Rohstoffverbrauch. IX, 181 Seiten. 1983.

Vol. 215: Semi-Infinite Programming and Applications. Proceedings, 1981. Edited by A.V. Fiacco and K.O. Kortanek. XI, 322 pages. 1983.

Vol. 216: H.H. Müller, Fiscal Policies in a General Equilibrium Model with Persistent Unemployment. VI, 92 pages. 1983.

Vol. 217: Ch. Grootaert, The Relation Between Final Demand and Income Distribution. XIV, 105 pages. 1983.

Vol. 218: P. van Loon, A Dynamic Theory of the Firm: Production, Finance and Investment. VII, 191 pages. 1983.

Vol. 219: E. van Damme, Refinements of the Nash Equilibrium Concept. VI, 151 pages. 1983.

Vol. 220: M. Aoki, Notes on Economic Time Series Analysis: System Theoretic Perspectives. IX, 249 pages. 1983.

Vol. 221: S. Nakamura, An Inter-Industry Translog Model of Prices and Technical Change for the West German Economy. XIV, 290 pages. 1984.

Vol. 222: P. Meier, Energy Systems Analysis for Developing Countries. VI, 344 pages. 1984.

Vol. 223: W. Trockel, Market Demand. VIII, 205 pages. 1984.

Vol. 224: M. Kiy, Ein disaggregiertes Prognosesystem für die Bundesrepublik Deutschland. XVIII, 276 Seiten. 1984.

Vol. 225: T.R. von Ungern-Sternberg, Zur Analyse von Märkten mit unvollständiger Nachfragerinformation. IX, 125 Seiten. 1984

Vol. 226: Selected Topics in Operations Research and Mathematical Economics. Proceedings, 1983. Edited by G. Hammer and D. Pallaschke. IX, 478 pages. 1984.

Vol. 227: Risk and Capital. Proceedings, 1983. Edited by G. Bamberg and K. Spremann. VII, 306 pages. 1984.

Vol. 228: Nonlinear Models of Fluctuating Growth. Proceedings, 1983. Edited by R.M. Goodwin, M. Krüger and A. Vercelli. XVII, 277 pages. 1984.

Vol. 229: Interactive Decision Analysis. Proceedings, 1983. Edited by M. Grauer and A.P. Wierzbicki. VIII, 269 pages. 1984.

Vol. 230: Macro-Economic Planning with Conflicting Goals. Proceedings, 1982. Edited by M. Despontin, P. Nijkamp and J. Spronk. VI, 297 pages. 1984.

Vol. 231: G.F. Newell, The M/M/∞ Service System with Ranked Servers in Heavy Traffic. XI, 126 pages. 1984.

Vol. 232: L. Bauwens, Bayesian Full Information Analysis of Simultaneous Equation Models Using Integration by Monte Carlo. VI, 114 pages. 1984.